AF890483

Hybrid Planar - 3D Waveguiding Technologies

Edited by Marcos D. Fernandez, José A. Ballesteros, Héctor Esteban and Ángel Belenguer

Published in London, United Kingdom

Hybrid Planar - 3D Waveguiding Technologies
http://dx.doi.org/10.5772/intechopen.100731
Edited by Marcos D. Fernandez, José A. Ballesteros, Héctor Esteban and Ángel Belenguer

Contributors
Angela Coves, Maurizio Bozzi, Zhongmao Li, Mengjie Qin, Pengzhan Liu, Xin Qiu, María García-Vigueras, Lucas Polo-López, Charalampos Stoumpos, Aurélie Dorlé, Raphaël Gillard, Carlos Molero, Abdel R. Sebak, Mohamed Mamdouh M. Ali, Islam Afifi, Mohammad Ali AbdElraheem, Isabelle Huynen, Vivien Van Kerckhoven, Luc Piraux, Yushi Chu, Jianzhong Zhang, Gang-Ding Peng, Liling Dong, Yanhua Luo, Darío Herraiz, José A. Ballesteros, Leticia Martínez, Marcos D. David Fernandez, Héctor Esteban, Ángel Belenguer

© The Editor(s) and the Author(s) 2023
The rights of the editor(s) and the author(s) have been asserted in accordance with the Copyright, Designs and Patents Act 1988. All rights to the book as a whole are reserved by INTECHOPEN LIMITED. The book as a whole (compilation) cannot be reproduced, distributed or used for commercial or non-commercial purposes without INTECHOPEN LIMITED's written permission. Enquiries concerning the use of the book should be directed to INTECHOPEN LIMITED rights and permissions department (permissions@intechopen.com).
Violations are liable to prosecution under the governing Copyright Law.

Individual chapters of this publication are distributed under the terms of the Creative Commons Attribution 3.0 Unported License which permits commercial use, distribution and reproduction of the individual chapters, provided the original author(s) and source publication are appropriately acknowledged. If so indicated, certain images may not be included under the Creative Commons license. In such cases users will need to obtain permission from the license holder to reproduce the material. More details and guidelines concerning content reuse and adaptation can be found at http://www.intechopen.com/copyright-policy.html.

Notice
Statements and opinions expressed in the chapters are these of the individual contributors and not necessarily those of the editors or publisher. No responsibility is accepted for the accuracy of information contained in the published chapters. The publisher assumes no responsibility for any damage or injury to persons or property arising out of the use of any materials, instructions, methods or ideas contained in the book.

First published in London, United Kingdom, 2023 by IntechOpen
IntechOpen is the global imprint of INTECHOPEN LIMITED, registered in England and Wales, registration number: 11086078, 5 Princes Gate Court, London, SW7 2QJ, United Kingdom

British Library Cataloguing-in-Publication Data
A catalogue record for this book is available from the British Library

Additional hard and PDF copies can be obtained from orders@intechopen.com

Hybrid Planar - 3D Waveguiding Technologies
Edited by Marcos D. Fernandez, José A. Ballesteros, Héctor Esteban and Ángel Belenguer
p. cm.
Print ISBN 978-1-80356-149-3
Online ISBN 978-1-80356-150-9
eBook (PDF) ISBN 978-1-80356-151-6

We are IntechOpen,
the world's leading publisher of
Open Access books
Built by scientists, for scientists

6,200+
Open access books available

169,000+
International authors and editors

185M+
Downloads

156
Countries delivered to

Our authors are among the
Top 1%
most cited scientists

12.2%
Contributors from top 500 universities

WEB OF SCIENCE™

Selection of our books indexed in the Book Citation Index
in Web of Science™ Core Collection (BKCI)

Interested in publishing with us?
Contact book.department@intechopen.com

Numbers displayed above are based on latest data collected.
For more information visit www.intechopen.com

Meet the editors

Marcos D. Fernandez received his degree in telecommunications engineering from the Universitat Politècnica de Catalunya (UPC), Spain in 1996, and his Ph.D. from the Universidad Politécnica de Madrid (UPM) in 2006. He joined the Universidad de Castilla-La Mancha in 2000, where he is now an associate professor in the Department of Electrical, Electronic, Automation and Communications Engineering. Since 2021, he has been Dean of the Escuela Politécnica de Cuenca. He has authored or co-authored several papers in peer-reviewed international journals and conference proceedings. His research interests are empty substrate-integrated waveguide (ESIW) devices, and their manufacture and applications.

José A. Ballesteros received his degree in telecommunications engineering from the Universidad de Alcalá de Henares (UAH), Spain in 2009, and his Ph.D. from the Universidad Politécnica de Madrid (UPM) in 2014. He joined the Universidad de Castilla-La Mancha in 2007, where he is now an associate professor in the Department of Electrical, Electronic, Automation and Communications Engineering. He has authored or co-authored several papers in peer-reviewed international journals and conference proceedings. His research interests are empty substrate-integrated waveguide (ESIW) devices, and their manufacture and applications.

Ángel Belenguer received his degree in telecommunications engineering from Universidad Politècnica de València (UPV), Spain in 2000, and his Ph.D., also from UPV, in 2009. He joined Universidad de Castilla-La Mancha in 2000, where he is now a professor in the Department of Electrical, Electronic, Automation and Communications Engineering. He has authored or co-authored more than 50 papers in peer-reviewed international journals and conference proceedings and frequently acts as a reviewer for several international technical publications. His research interests include methods in the frequency domain for the full-wave analysis of open-space and guided multiple scattering problems, the application of accelerated solvers or solving strategies to new problems or structures, EM metamaterials, and empty substrate-integrated waveguide (ESIW) devices and applications.

Héctor Esteban received his degree in telecommunications engineering and his Ph.D. from the Universidad Politècnica de València (UPV), Spain, in 1996 and 2002, respectively. He worked with the Joint Research Centre of the European Commission in Ispra, Italy and with the European Topic Centre on Soil (European Environment Agency). In 1998 he rejoined the UPV, where he is a professor and Dean of the School of Telecommunications Engineering. His research interests include methods for the full-wave analysis of open-space and guided multiple scattering problems, and CAD design of microwave devices, especially using new empty substrate-integrated waveguide technologies, and its characterization for use in space conditions.

Contents

Preface

A rectangular waveguide (RWG) is the best technology to guide high-frequency signals in traditional high-performance communication systems. As is well known, it is characterized by low losses, high power handling capacity and mechanical robustness. Its main drawbacks are size, weight and cost, which invalidate its use in new and emerging communication systems, such as constellations of low-cost and low-Earth-orbit satellites.

The most popular alternative to RWG is usually printed circuit technology, such as microstrip or coplanar lines. Despite the benefits of reduced cost, volume and ease of manufacture, these technologies present high power losses.

Since 2001, new planar and substrate-integrated waveguides have been developed that approach the desirable performance of an RWG, but are synthesized on one or more printed circuit boards. The first was the substrate-integrated waveguide (SIW), which emulates a dielectric-filled RWG in a single circuit board where the side walls are made of metallic via holes. Although SIW is a good alternative to classic planar technologies, the presence of lossy dielectric makes it impossible to achieve performance similar to that of the RWG. In 2014, the empty substrate-integrated waveguide (ESIW) was introduced as a composite of three soldered metalized circuit board layers where the middle layer had been emptied to emulate an RWG. ESIW is now the best approach to an RWG in terms of performance, but retains the characteristics of planar circuits: ease, compactness, mass production, low volume, low weight and low cost. Among other alternatives are the air-filled substrate-integrated waveguide (AFSIW), which is the practical implementation of the modified substrate-integrated waveguide (MSIW); the hollow substrate-integrated waveguide (HSIW), similar to AFSIW but implemented in low-temperature cofired ceramic (LTCC) technology; and the partially dielectric-filled empty substrate-integrated waveguide (PFESIW), which is an ESIW loaded with a longitudinal dielectric slab. Newer hybrid planar–3D waveguiding structures have arisen, implementing waveguides of either a single conductor (no TEM mode) or two conductors (pure TEM mode).

These novel hybrid technologies are receiving much research attention. Among the studies that have been published are several devices, advances in topology variations and in the manufacturing process, and newer techniques combining traditional planar processes with metal and plastic 3D printing. The maturity of these technologies and their use by the communications industry may increase the performance of communication devices, which in turn will have a major economic impact on the high-frequency communication sector.

The goals of this book are to present the basis of these new hybrid structures and to introduce advances in the design of devices and systems, manufacturing processes and tests, as well as the various applications where these technologies can be used.

The academic editorial board of this book wishes to thank all the colleagues whose contributions have made this project possible.

Marcos D. Fernandez, José A. Ballesteros and Ángel Belenguer
Escuela Politécnica de Cuenca,
Universidad de Castilla-La Mancha,
Cuenca, Spain

Héctor Esteban
Universitat Politècnica de València,
València, Spain

Section 1

Traditional Manufacturing

Chapter 1

SIW-Based Devices

Zhongmao Li, Mengjie Qin, Pengzhan Liu and Xin Qiu

Abstract

With the development of microwave wireless communication technology, microwave passive frequency selectivity circuits are developing toward multi-function and miniaturization. Substrate-integrated waveguide (SIW) devices have good development prospects in miniaturization and integration. The research on devices based on SIW is lack unified theoretical combing. This chapter describes the novel circuits based on substrate-integrated waveguide, including filter, power divider, switch, phase shifter, diplexer, and crossover. This chapter describes the design methods and special properties of these circuits. The display of these circuits is expected to give some inspiration to your research.

Keywords: substrate-integrated waveguide, filter, power divider, switch, phase shifter, diplexer, crossover

1. Introduction

Microwave devices are generally designed using metallic waveguides and planar technology. Traditional metallic waveguide is low insertion loss (IL), and high-quality factor (Q value), but the components are bulky and nonplanar, which cannot be integrated into RF integrated circuits. Planar technology (microstrip transmission lines or coplanar waveguides) could be easily integrated, but the insertion loss and Q value are worse. At higher frequencies (>30 GHz), planar structures are prevented to apply due to high transmission losses [1].

To tackle the above-mentioned problems, the concepts of SIW are proposed [2, 3]. The SIW is considered a quasi-waveguide, developed by two parallel copper sheets with rows of conducting cylinders embedded in a dielectric substrate on each side to connect the two plates, as shown in **Figure 1**. When the parameters of the design conform following rules, SIW can be equivalent to a conventional rectangular waveguide [4].

$$s > d \quad (1)$$

$$s/\lambda_c < 0.25 \quad (2)$$

$$l/k_o < 1 \times 10^{-4} \quad (3)$$

$$s/\lambda_c > 0.05 \quad (4)$$

Where s is the gap between the adjacent vias, d is the diameter of a via, λ_c is the cutoff wavelength, l is the total loss, and k_o is the wave number in a vacuum.

IntechOpen

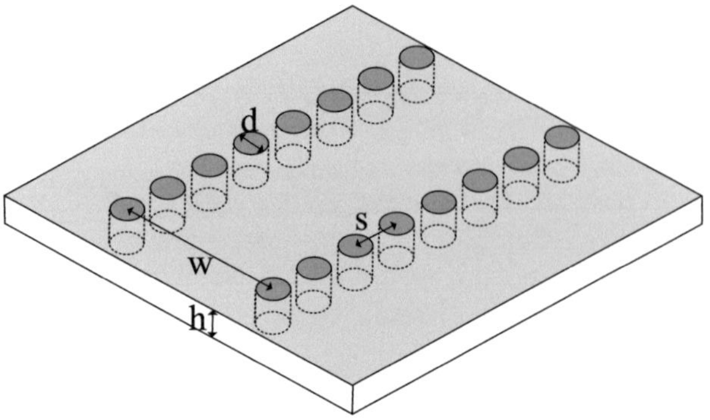

Figure 1.
Three-dimensional view of SIW.

In this way, a rectangular metallic waveguide filled with a dielectric material is constructed in planar form, the most key advantage, thus making a complete integration with other planar transmission-line circuits on the same substrate. Meanwhile, SIW structures maintain most of the benefits of classical metallic waveguides. Thus, SIW is a suitable choice for less lossy, compact, and simple microwave and mm-wave systems.

After 20 years of development, SIW has been used to design all kinds of passive circuits, such as filters, power dividers, switches, phase shifters, diplexers, and cross-overs, and presents high performance. At the same time, many techniques have been developed in recent times. This chapter provides a brief, yet necessary, understanding of current research on the SIW components.

2. SIW filters

Traditional planar transmission structures, such as microstrip and coplanar waveguide (CPW), have been widely used in filter design. Compared with these two structures, the substrate-integrated waveguide technology has the advantages of high Q coefficient, low cost, and high integration, and is widely used in the design of narrowband and broadband band-pass filters. The substrate-integrated waveguide technology can realize a variety of filters with different characteristics, including multi-band filters, wide stop-band filters, and reconfigurable filters. In the early days, SIW filters focused on the realization of circuit functions, and reports on the miniaturization of SIW filters appeared around 2005 [5]. There are various miniaturization technologies for SIW filters, and miniaturization has become the main trend in the development of SIW filters. This section introduces the application of SIW technology in filters from five aspects.

2.1 Conventional substrate-integrated waveguide filters

The design of the SIW bandpass filter can follow a similar approach to the design of gas-filled waveguide filters, or it can be based on the coupled matrix approach. In [6], a three-pole Chebyshev filter is designed through the inductive column synthesis technology, and its insertion loss and return loss are more than 1 dB and 17 dB, respectively. This SIW-based filter reduces size, weight, and cost significantly.

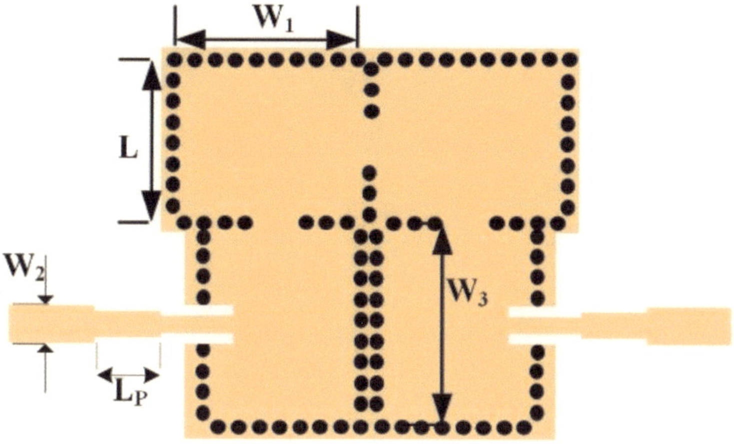

Figure 2.
Configuration of proposed SIW filter [7].

A novel wide-stopband SIW filter using an angular cavity is performed in Ref. [7]. For each corner cavity, two transmission zero points can be introduced, and the position of each transmission zero point can be easily controlled by adjusting the appropriate geometric dimensions. **Figure 2** shows the configuration of the proposed SIW filter.

2.2 Multilayer substrate-integrated waveguide filters

The emergence of multi-layer technology can extend the traditional two-dimensional planar structure to a three-dimensional direction, greatly increasing the flexibility and freedom of filter design, and can increase the coupling mode. At present, the realization process of multi-layer structure mainly includes low temperature cofired ceramic (LTCC) process and multi-layer PCB process. The second part mainly introduces the multi-layer structure design of the filter applied to the substrate-integrated waveguide.

The first multilayer technology is the substrate-integrated-folded waveguide (SIFW), which was proposed by Grigoropoulos et al. [8]. The use of substrate-integrated folded waveguides in filter design can greatly reduce the overall circuit size of the filter.

Figure 3 shows a filter based on a SIFW resonator design [9]. The filter was developed using LTCC technology to achieve miniaturization. Resonators of the same layer, such as resonator 1 and resonator 4, resonator 2 and resonator 3, are horizontally coupled through the corresponding sensing window of the through hole. The different layers of resonators, such as Resonator 1 and Resonator 2, Resonator 3 and Resonator 4, are vertically coupled through a slot in the common wall between the two layers. Compared with the traditional planar direct coupled waveguide filter, the area of the filter can be reduced to 26.3%.

The second multilayer technology is the ridged substrate-integrated waveguide (RSIW). As shown in **Figure 4** [10], RSIW is the introduction of a longitudinal metal ridge into a classical waveguide without any degradation in RF performance. At the same time, the use of ridge waveguides can increase the bandwidth by 37%.

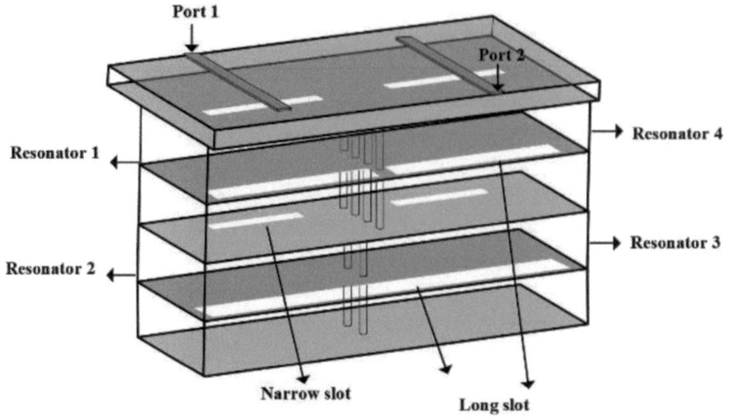

Figure 3.
Configuration of proposed SIFW filter [9].

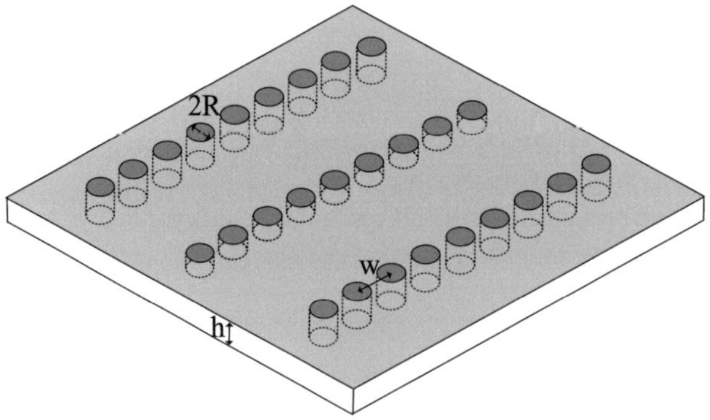

Figure 4.
The dimensional structure of the RSIW [10].

2.3 Fractional mode substrate-integrated waveguide filters

The electric field distributions in the conventional SIW, half-mode SIW (HMSIW), quarter-mode SIW (QMSIW), eighth-mode SIW (EMSIW), sixteenth-mode SIW (SMSIW), and 32-mode SIW (TMSIW) cavities at the dominant (TE101) mode are shown in **Figure 5**. The symmetry planes A-A1, B-B1, C-C1, and D-D1 are considered to be magnetic walls, and the remaining planes are electric walls. The full-mode SIW cavity can be cut into half-modes by cutting along the symmetry line B-B1, as shown in **Figure 5b**. It is obvious that the fundamental mode remains unchanged and the volume is reduced by half. The HMSIW cavity can be cut into quarter modes along the O-A1 line, as shown in **Figure 5c**. Compared with HMSIW, QMSIW achieves a 50% reduction. The QMSIW can be reduced to the EMSIW by cutting the QMSIW cavity along the O-D1 line, as shown in **Figure 5d**. Similarly, the electric field distribution is unchanged. SMSIW and TMSIW can be obtained by cutting the EMSIW cavity along the O-E line and O-F line, respectively, as shown in **Figure 5e** and **f**. The resonant frequency and electric field distribution of all sub-modes remain the same.

It is worth noting that since the topology of the fractional mode technique is inherently open-structured, there will be undesired radiation leakage, which can

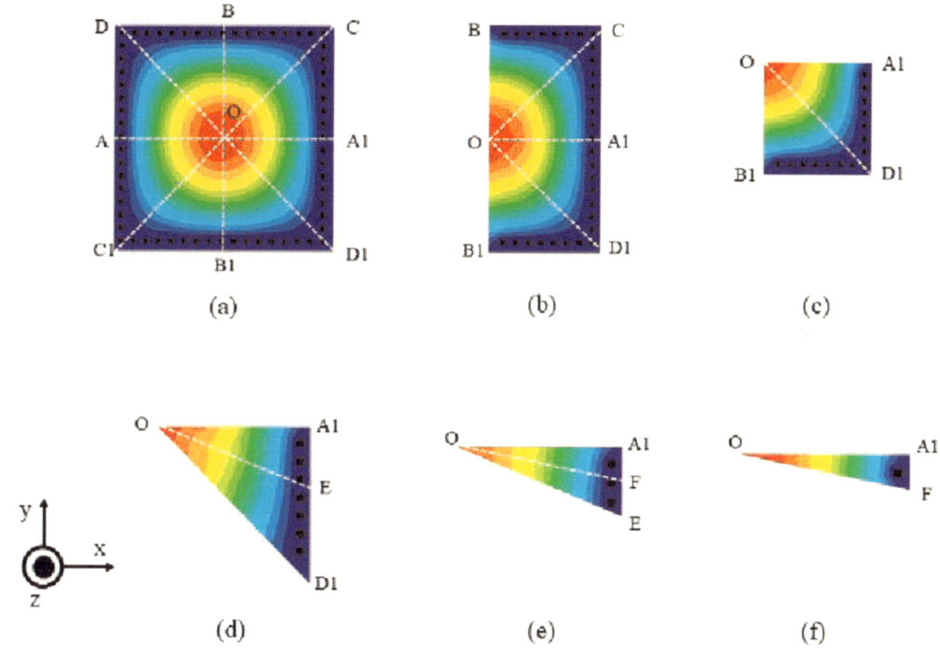

Figure 5.
Electric field distributions of (a) full-mode SIW, (b) HMSIW, (c) QMSIW, (d) EMSIW, (e) SMSIW, (f) TMSIW [11].

Figure 6.
Electric field distribution for the fundamental mode of the SIW resonators. (a) Conventional QMSIW cavity, (b) shielded QMSIW cavity [12].

sometimes significantly reduce the quality factor. To solve this problem, some methods are applied. Taking quarter mode as an example, the shielded QMSIW is used to design the filter in Ref. [12]. As shown in **Figure 6**, the shielded quarter-mode resonator is made by placing two rows of metal through holes in the opening, which are called shield walls. Adding shielding walls can partially block the propagation of waves along the surface, reducing the efficiency of the radiating side of the structure. Compared with the conventional QMSIW, the shielded QMSIW exhibits better performance.

2.3.1 Half-mode SIW (HMSIW) filters

The half-mode substrate-integrated waveguide was proposed by Wei Hong et al. in 2006 [13]. This waveguide structure is obtained by cutting the SIW so that the symmetry plane along the transmission direction is equivalent to the magnetic wall.

As shown in **Figure** 7, the HMSIW is 50% smaller in size while keeping the electric field distribution unchanged. Since the volume of HMSIW is reduced by half and the propagation characteristics remain unchanged, it has attracted the attention of many scholars.

A compact dual-band filter based on HMSIW is shown in **Figure 8** [14]. The dual-band filter is designed by the quasi-TEM mode and the TE102 mode of the novel HMSIW resonator. The resonant frequency of TE102 can be adjusted by the slot line, which has little effect on the performance of the quasi-TEM mode. To improve the out-of-band rejection, the source negative coupling structure is applied. The measurement results show that the first and second passbands are mainly concentrated at 2.41 and 3.51 GHz, the relative bandwidths are 10.8% and 6.4%, respectively, and the minimum insertion losses of the two passbands are 1.45 and 1.74 dB, respectively.

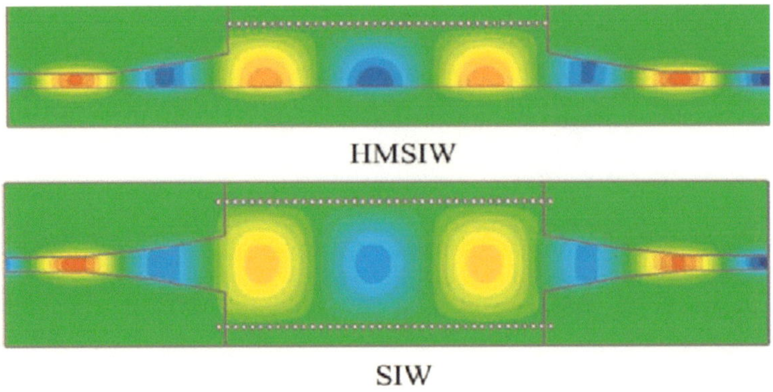

Figure 7.
Dominant field distribution in HMSIW and SIW [13].

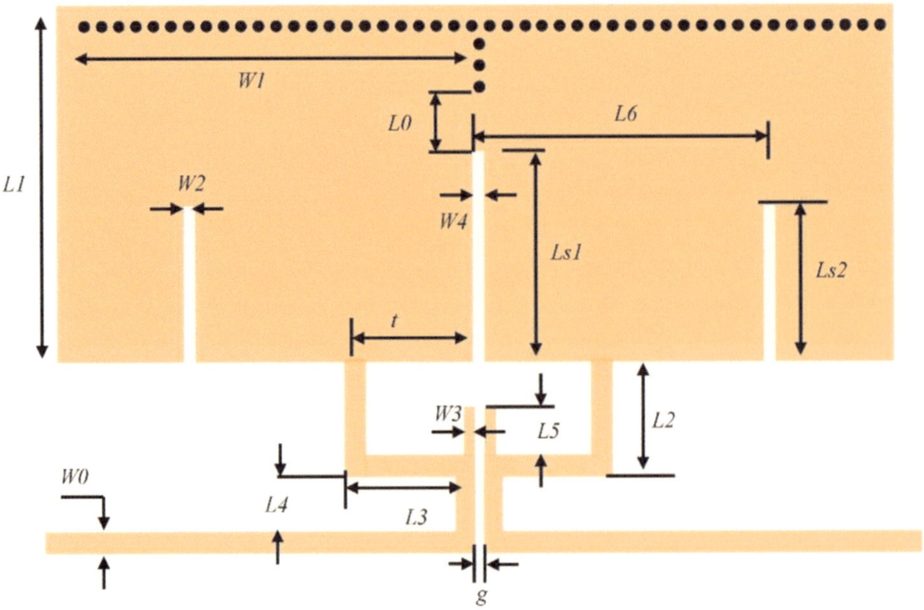

Figure 8.
Configuration of the proposed dual-band filter [14].

2.3.2 Quarter-mode SIW (QMSIW) filters

The authors of Ref. [15] proposed the bandpass filter based on QMSIW for the first time. The geometries of the two filters are shown in **Figure 9**. The bandpass filter consists of cascaded QMSIW cavities, and its overall size is greatly reduced. For verification, two filters are designed. The prototype I is designed at the center frequency of 5.85 GHz with a fractional bandwidth of about 14%. Prototype II is designed at the center frequency of 5.5 GHz with a fractional bandwidth of about 26%.

The authors of Ref. [16] reported a QMSIW for designing a series of bandpass filters. Quarter-mode SIW cavities can be connected in various ways, such as sharing a side or sharing a corner, with some metal vias removed to allow cavity coupling. The authors introduce novel filter topologies based on the proposed two different techniques, side-coupling, and corner-coupling. Taking the filter shown in **Figure 10** as an

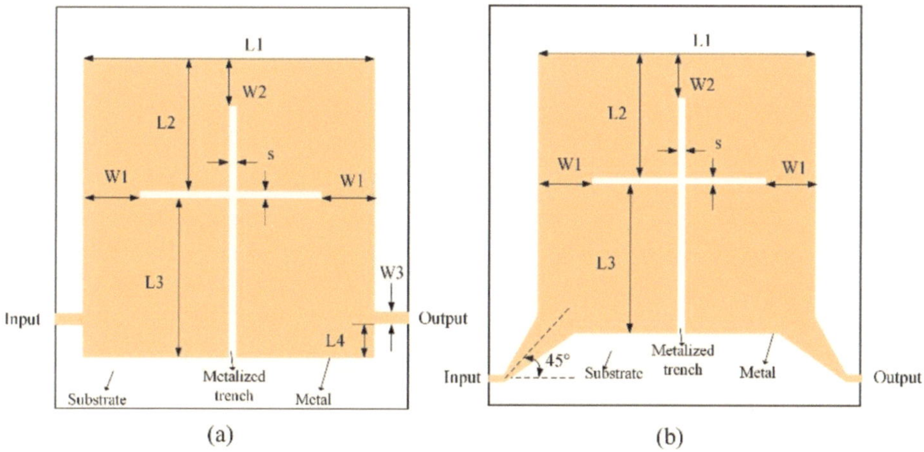

Figure 9.
The geometries of the two filters. (a) The prototype I, (b) the prototype II [15].

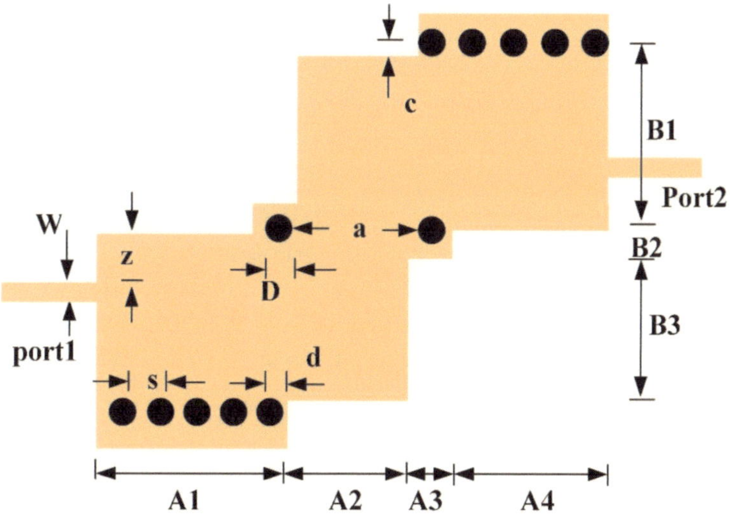

Figure 10.
Configuration of the side-coupled four-pole filter [16].

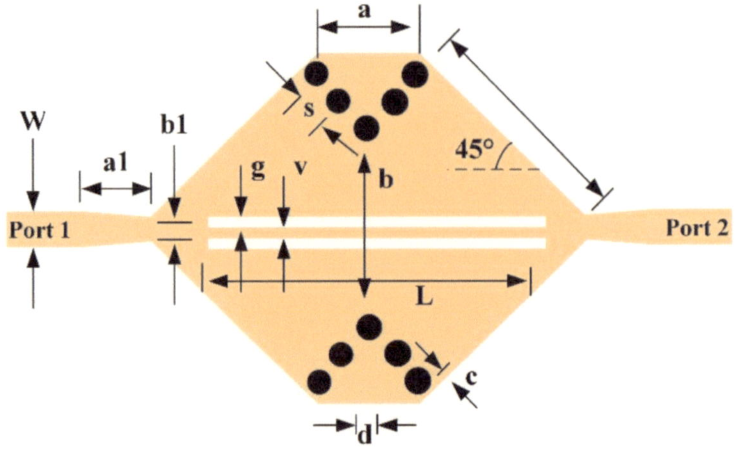

Figure 11.
Configuration of the corner-coupled four-pole filter [16].

example, the resonator is laterally coupled by removing some through holes in the common wall. By using a plus-single topology, it can be used to implement filters with any number of poles. **Figure 11** shows the filter composed of the new coupling. By using a coplanar resonator and quarter-mode cavity together, a three-pole filter with a center frequency of 4 GHz is designed, and its relative bandwidth is 16%.

2.3.3 Eighth-mode SIW (EMSIW) filters

The QMSIW can be further bisected into two parts along the symmetrical plane with one electric wall and two magnetic walls forming the EMSIW.

The authors of Ref. [17] reported a systematic research campaign on the EMSIW filter. The authors conducted a series of studies on eighth-mode filters. Compared with the traditional SIW cavity, the one-eighth mold cavity has a compact size. This paper first explores the four feeding modes of the eighth mode, then designs a variety of bandpass filters for the eighth mode utilizing electrical coupling and magnetic coupling, and finally designs two bandpass filters. The first filter takes the form of a combination of an eighth mode and a coplanar waveguide. Through the mutual coupling between the TE101 mode of the two EMSIW cavities and the quasi-TEM mode of the folded coplanar waveguide, a band-pass filter with a center frequency of 9.1 GHz is finally designed. The second filter adopts a multi-layer process, and the upper and lower substrates are coupled through two rectangular slots. Compared with filter 1, this filter has more transmission zeros. The proposed EMSIW BPFs have the merits of being compact in size and high selectivity. The geometry of the two filters is shown in **Figures 12** and **13**.

2.3.4 High-mode SIW (SMSIW and TMSIW) filters

A miniaturized bipolar bandpass filter based on a SMSIW was investigated [18]. The geometry of the filter is shown in **Figure 14**. The filter operates in fundamental mode TM01. Under the same resonant frequency, the size of the SMSIW cavity is only 1/16 of that of the traditional SIW cavity. To suppress the higher-order mode TM02, a rectangular slot is etched in the ground plane below the transmission line. The overall size of the filter is compact and has good out-of-band performance.

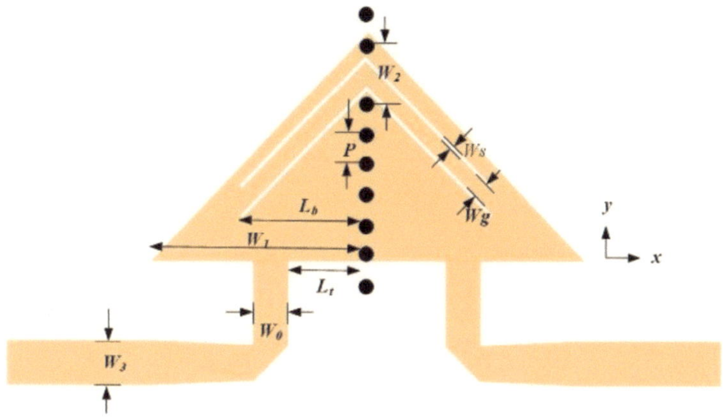

Figure 12.
Geometry of the triple-order EMSIW BPF with CPW [17].

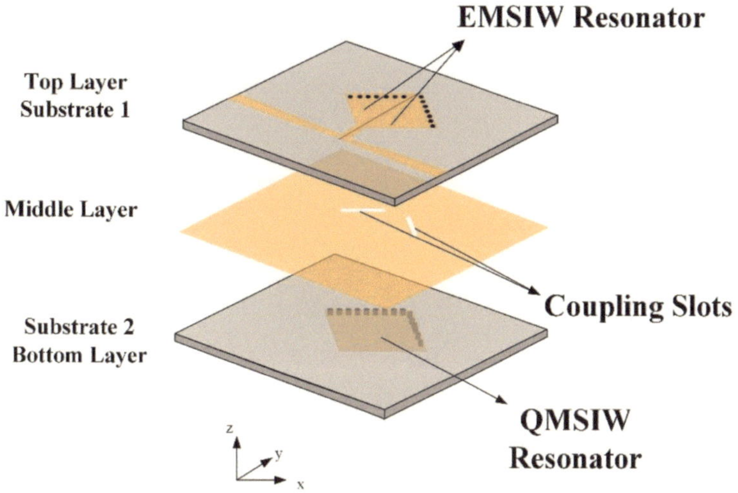

Figure 13.
Geometry of the triple-order EMSIW BPF with QMSIW cavity [17].

A TMSIW bandpass filter is presented in Ref. [19]. The geometry of the filter is shown in **Figure 15**. By adjusting the position of the feed and the coupling distance, the frequency position of the passband can be selectively achieved. The TM0n0 mode has a good unloaded quality factor in addition to providing different frequency options. In addition, the perturbation of the arc slot can effectively tune the coupling and achieve high selectivity.

2.4 SIW filters loaded with complementary Split-ring resonators (CSRR)

As an electromagnetic metamaterial, CSRR is often used in the miniaturization of filters. CSRR is an electrical resonator that can be excited by an axial electric field, which is proposed by Pendry [20]. In Ref. [21], two HMSIW filters loaded with the stepped-impedance CSRRs are presented. The geometries of the two filters are shown

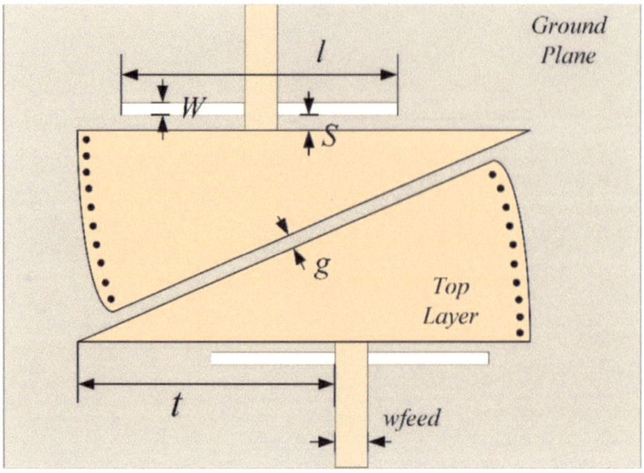

Figure 14.
Geometry of the proposed filter [18].

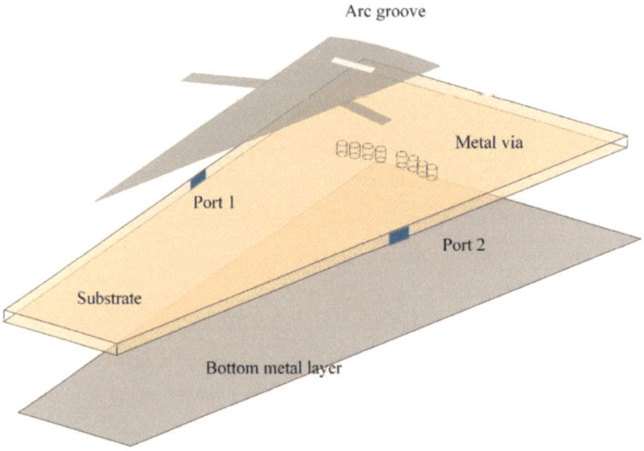

Figure 15.
Geometry of the triple-order EMSIW BPF with QMSIW cavity [19].

in **Figure 16**. By loading both CSRR and stepped-impedance complementary split-ring resonator (SICSRR) into the two sides of the HMSIW cavity, a dual-band resonator is formed. By adjusting the size of CSRR and SICSRR, the position of the two passbands can be flexibly adjusted. Compared with other dual-band filters, the HMSIW-SICSRR filter size is reduced by 48.5% and 50%, respectively.

Novel bandpass filters are reported by using HMSIW loaded with CSRR and a capacitive metal patch [22]. The geometry of the proposed filter is shown in **Figure 17**. Based on this combined structure, independent double passbands can be generated. In addition, the external quality factor and coupling factor of the two frequency bands can be adjusted independently.

2.5 Tunable SIW filters

A new method to design tunable filters is shown in **Figure 18** [23]. The proposed filter consists of a conventional SIW resonator and an additional via. The vial contains an open-loop slot in the top metal wall, where the size, location, and orientation of the

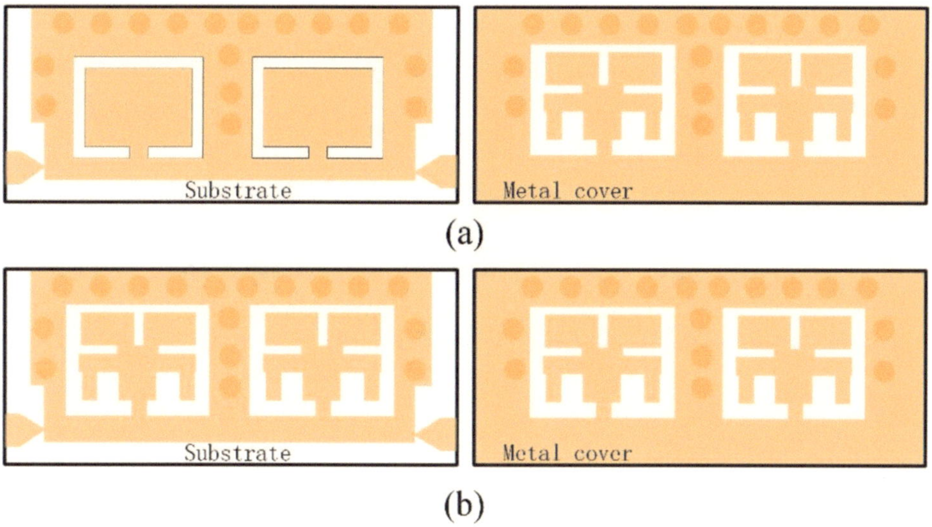

Figure 16.
Geometry of the proposed filters. (a) Loaded with CSRR and SICSRR, (b) loaded with SICSRRs [21].

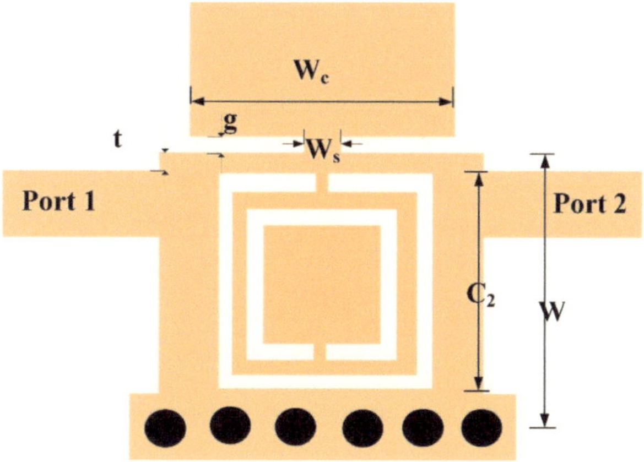

Figure 17.
Geometry of the triple-order EMSIW BPF with QMSIW cavity [22].

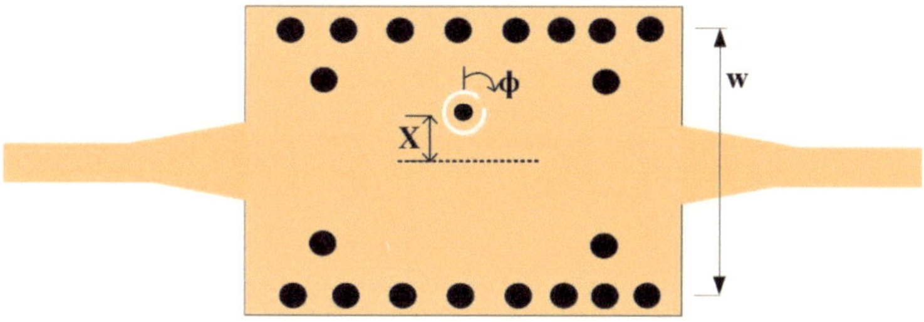

Figure 18.
Geometry of the proposed filter [23].

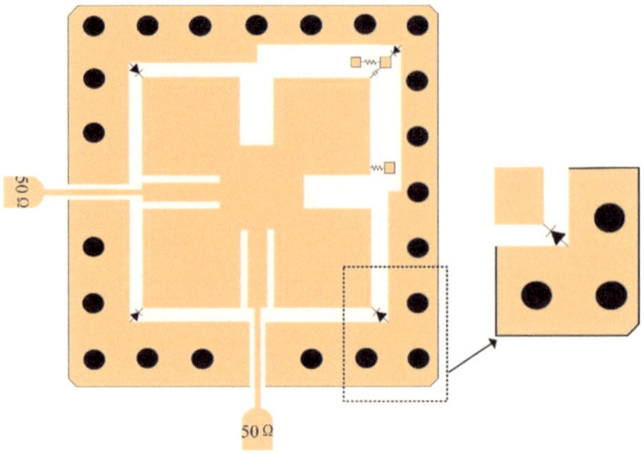

Figure 19.
Geometry of the proposed filter [24].

open-loop slot determine the current and field distribution paths, which in turn controls the tuning range, resulting in an 8% tuning range.

A tunable filter based on microstrip patch resonators is proposed [24]. The overall layout of the filter is shown in **Figure 19**. The designed filter is similar to the SIW structure, that is, an electrical wall composed of through-holes is placed around the microstrip patch to reduce radiation, thereby reducing the insertion loss of the filter. In addition, the varactor diode is selected as the tuning element because of its small size, fast tuning speed, and large tuning range. The final designed filter covers six different states from 2.0 GHz to 2.53 GHz, and its insertion loss is 1.68 dB–3.90 dB.

Table 1 shows the performance parameter of the filters following different technological approaches.

Device	Technology	Size ($\lambda_g * \lambda_g$)	Center frequency (GHz)	FBW (%)	Insertion loss (dB)	Return loss (dB)
[7]	SIW	—	20	27.5	2.78	16
[9]	SIFW	0.3*0.88	30.2	11.3	3.7	13.5
[14]	HMSIW	0.45*0.67	2.41/3.51	10.8/6.4	1.45/1.74	20/20
[15]	QMSIW	0.5*0.5	5.85	14	2	13
[15]	QMSIW	0.5*0.5	5.5	26	1.2	13
[17]	EMSIW-CPW	0.33*0.33	9.1	22	0.9	23
[17]	EMSIW-Multilayer	0.32*0.32	9.1	19.8	1.3	20
[18]	SMSIW	—	2.05	7	1.2	16
[19]	TMSIW	0.468*0.198	7.26	6.8	1.5	—
[21]	HMSIW+ SICSRR	0.046*0.046	5.51/8.83	9.4/5	1.3/1.8	18/18
[21]	HMSIW+ SICSRR	0.045*0.045	5.52/8.81	9/5.4	1.3/2	23/14

Table 1.
Comparison of different filters.

3. SIW power divider

Power divider (PD) can be used in phased array antennas, multiplexers, power amplifiers, and so on. Substrate-integrated waveguides have attracted attention in the design of power dividers due to their strong anti-interference, smaller radiation loss, and higher quality factor (Q value). There are three main research directions on SIW power dividers in recent years. Firstly, SIW PDs are developed with higher isolation and lower insertion loss. The main isolation networks include one or more resistors and the microstrip-based network [25, 26]. Secondly, the filtering power dividers (FPD) are proposed to combine the power divider and filter to realize the miniaturization of the circuit [27–29]. Thirdly, ultra-wideband and wideband isolation PDs are fabricated [30, 31]. The main technical parameters of the power divider include insertion loss, return loss, isolation between ports, amplitude balance and phase balance, etc.

3.1 PD with high isolation

Wilkinson PD is used to achieve a high degree of isolation between output ports. An architecture of Wilkinson PD is based on fixed-width SIW lines with isolation resistors as shown in **Figure 20** [25]. The resistors are formed of various surface-mounted parallel resistors to reduce the effect of SIW-resistance discontinuity. The present topology shows enhanced isolation and adaptation in a wide bandwidth.

Another SIW PD with high output isolation is shown in **Figure 21** [26]. The power divider is constructed by using two substrate layers and three metal layers. Five microstrip lines are placed on the top metal layer and coupled with the SIW through three-slot lines located on the middle metal layer (common ground). Three resistors are used to improve the output isolation. This device has input return loss and

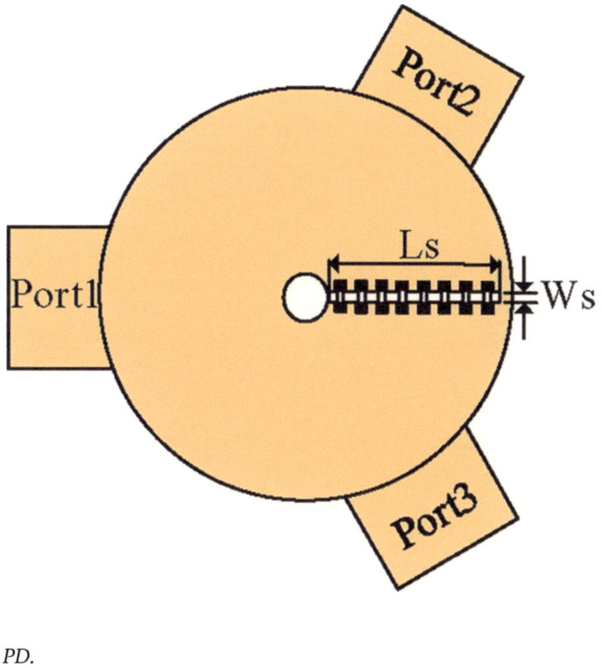

Figure 20.
The novel Wilkinson PD.

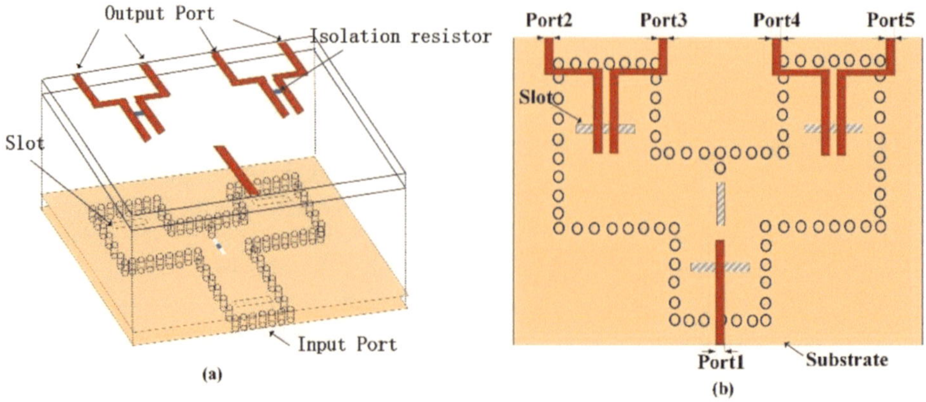

Figure 21.
Structure of proposed PD (a) 3D view. (b) Top view [26].

insertion loss greater than 17 dB and 1.6 dB, respectively. However, this structure increases the complexity of the process.

3.2 Filtering power divider

Loading a resonator or multiple resonators on the power divider can realize the dual function of filtering and power division. **Figure 22** shows a triple-band FPD and a quad-band FPD [27]. Two unequal CSRRs are engraved face to face on the top of the SIW rectangular cavity to achieve two operating frequencies. An outer pair of U-shaped slots (USS) or two pairs of USS is engraved on the bottom of the SIW to achieve the third or the fourth operating frequency. The fabricated FPD presents good isolation and return loss.

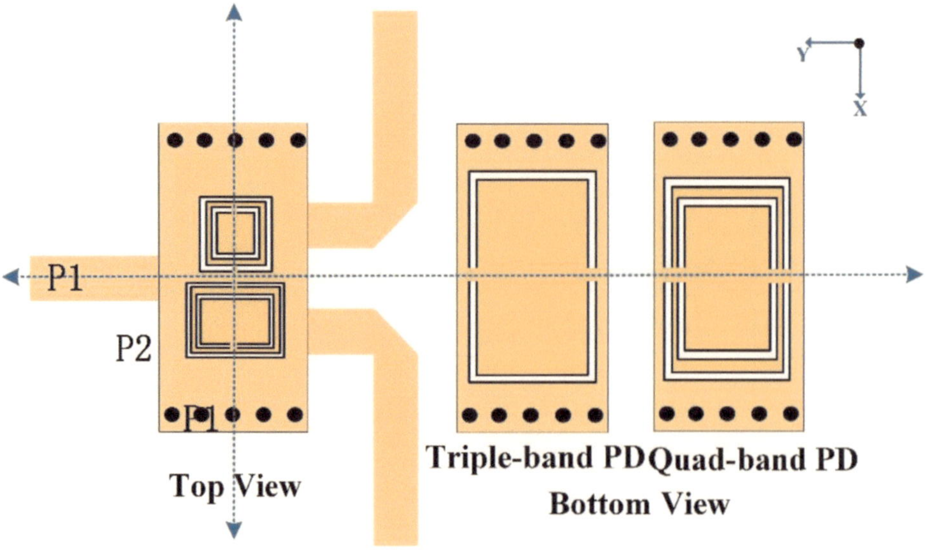

Figure 22.
Prototypes of Triple-band and quad-band PDs [27].

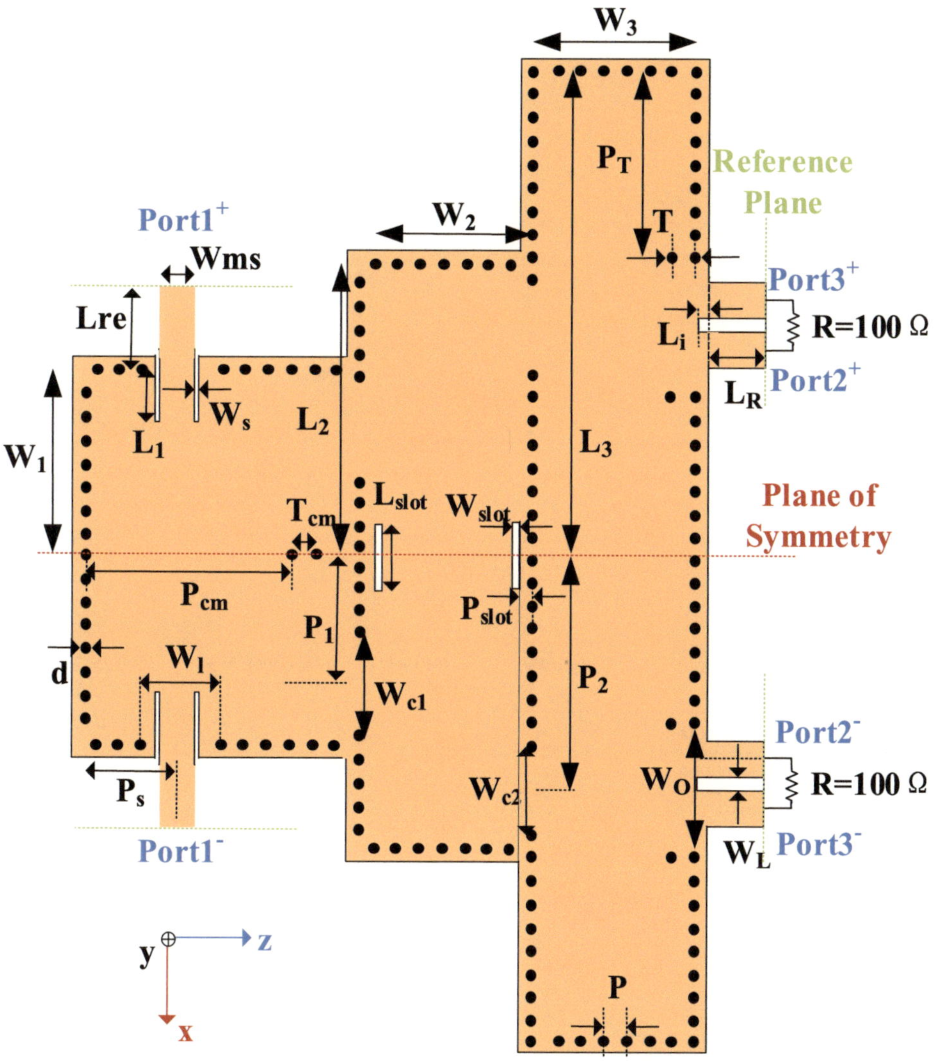

Figure 23.
Fabricated dual-band balanced SIW FPD.

Compared to the above FPD, the balanced SIW FPD is relatively less explored, especially for dual-band applications. **Figure 23** shows a single-layer dual-band SIW FPD which has three bisected substrate-integrated cavities [28]. By adjusting the position of metal vias, each cavity is designed to not only form three-pole dual passbands but also attain high in-band common-mode (CM) rejection. An isolation resistor is loaded across each output pair to attain high isolation easily.

3.3 Ultra-wideband and wideband isolation PD

Another filtering method is to load the slot line. As shown in **Figure 24**, two-slot lines are loaded on the SIW to get a third-order filtering response [30]. The isolation network consisting of the coplanar waveguide feeding line of Port 1, the interdigital capacitor, and the resistor is embedded in SIW, which expands the bandwidth of

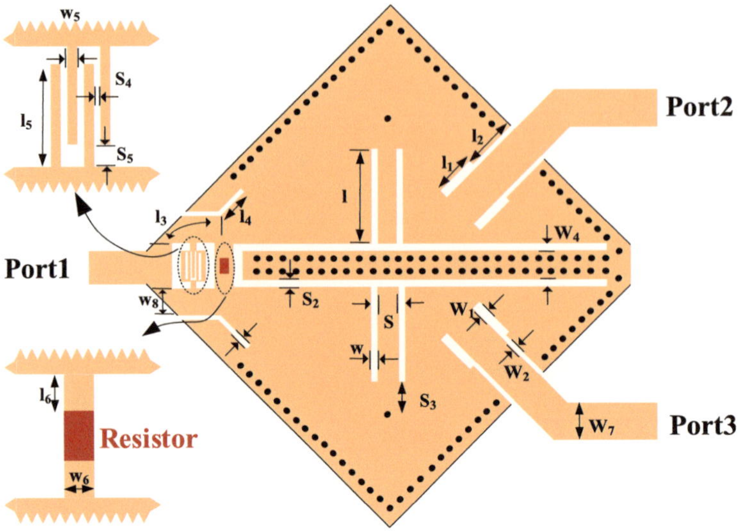

Figure 24.
Proposed SIW FPD [30].

Device	Technology	Bandwidth (GHz)	Isolation (dB)	Insertion loss (dB)	Return loss (dB)
[25]	Isolation network	8.85–10.6	20	3.3	15
[26]	Isolation network	9.2–10.8	20	1.6	17
[27]	FPD	3.47/4.73/6.31 3.34/4.82/6.17/7.66	14 12	3.8/4.85/4.06 3.5/3.95/4.26/4.85	20 19
[28]	FPD	28/39	20/14.9	1.7/2.2	—
[29]	FPD	10.64–11.84/ 13.37–13.97	16.1	1.4/2.2	17.8/ 18.4
[30]	Wideband isolation	5.59–6.4	21	1.05	20

Table 2.
Comparisons with different power dividers.

isolation and achieves lower insertion loss and good isolation. The proposed SIW FPD has isolation higher than 21 dB from 5.59 to 6.4 GHz.

Table 2 shows the performance parameter of the power dividers following different technological approaches.

4. SIW switch

The working principle of the SIW switch is to control the transmission of waves in the needed SIW path by various control methods, mainly including electronic control switches, magnetic control switches, and mechanical switches. Because of the ease of integration and low cost, PIN diodes are used as tunable devices loading on the surface of SIW for most electronic control switches. There have been many examples of designing switches using the difference of characters in diode forward and reverse bias [32–35]. Magnetic control switches are mostly loaded with ferrite [36, 37]. The permeability of the ferrite slabs loaded in the SIW cavity is tuned by an external transverse magnetic field, so that the cutoff frequency of the SIW mode is changed and the switching function is realized. Magnetic control switches have higher power handling capability and longer service life. The mechanical switch has the deficiencies of a short operation lifetime and slow switching speed [38]. However, it can avoid the problem of parasitism. A switch mainly focuses on three factors: insertion loss, return loss, and isolation.

4.1 Electronic control switch

An electronic control single-pole single-throw (SPST) switch is shown in **Figure 25**. Firstly, make a narrow slot in the SIW top wall, then, the diodes are capacitively coupled to the slot by metal pads on a very thin insulating layer. With the pin diodes being switched between on-state and off-state, the propagation could be transformed between two different types of modes. When the diodes are turned on, the top slot is closed and the waveguide becomes an ordinary SIW. When the diodes turn off, the electric field and propagation characteristics of the structure can be seemed as HMSIW, as long as the slot has an appropriate size [32]. The difference in cut-off frequency between SIW and HMSIW gives a switching bandwidth from 3 to 4.7 GHz. The fabricated switch is presented with isolation of 50 dB and 3 dB IL in 3–3.5 GHz.

In addition, PIN diodes are also used in a single pole double throw (SPDT) switch. Inductive posts with rectangular slots are embedded in the SIW to increase the impedance of the SIW, which can disturb the incident signal flow. The PIN diodes and bias networks are placed to control the operating mode of the inductive posts, as shown in **Figure 26**. The proposed SPDT switch has measured isolation(S31) of greater than 10 dB, isolation(S32) of greater than 15 dB, and an IL of less than 2.55 dB from 8.24 to 10.36 GHz [33]. However, SIWs are narrow at higher frequency ranges and cannot use switched vias. Thus, this method is no longer suitable at higher millimeter-wave frequencies.

Different from the above internal through-hole design, a SIW switch with a resonant slot is made innovatively [34]. The proposed switch in **Figure 27** has an IL of less than 1.3 dB and isolation of more than 10 dB in the switchable frequency band. The

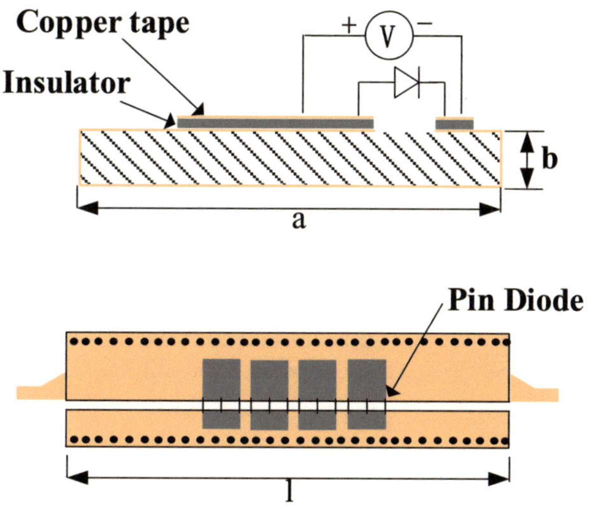

Figure 25.
Structure of switched waveguide [32].

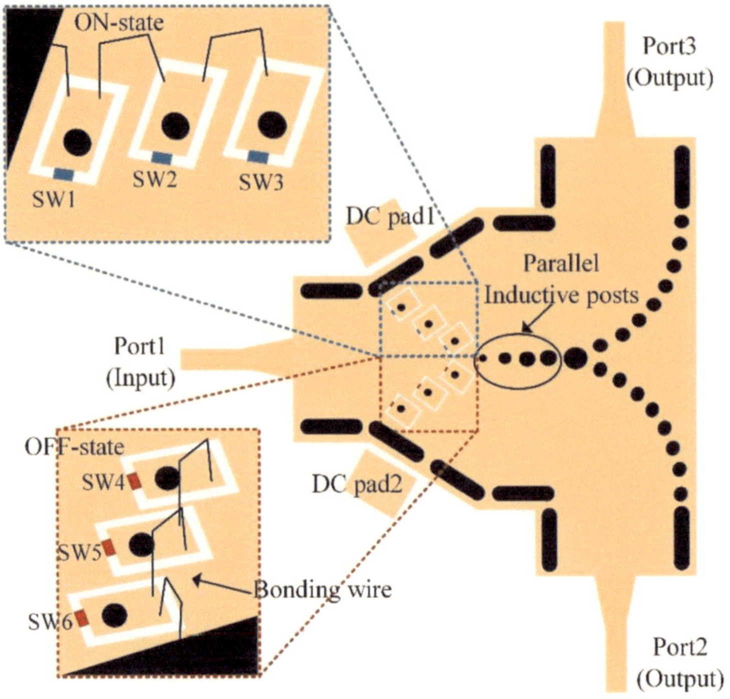

Figure 26.
Structure of the SIW SPDT switch [33].

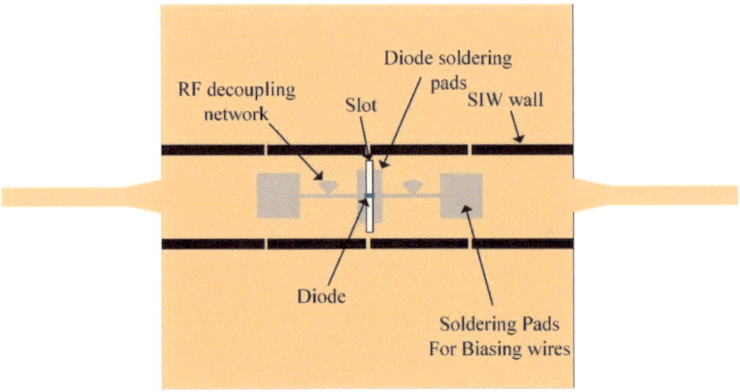

Figure 27.
A switch with a resonant slot [34].

resonant slot vanishes when the PIN diode is in the forward-biased state, while the resonant slot presents a high impedance load on the SIW when the PIN diode is in the reverse-biased. This novel device can be particularly suitable for higher millimeter-wave frequency due to its simple structure.

4.2 Magnetic control switch

There are many attempts to magnetically control using ferrite loading. A way to load is shown in **Figure 28**, which consists of rectangular ferrite slabs loaded on the

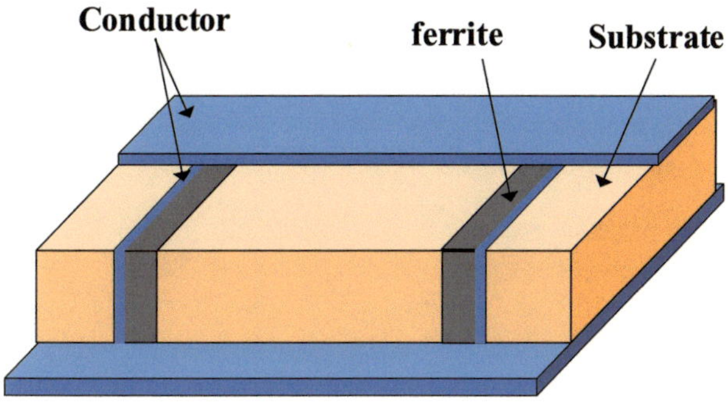

Figure 28.
Ferrite-loaded SIW switch.

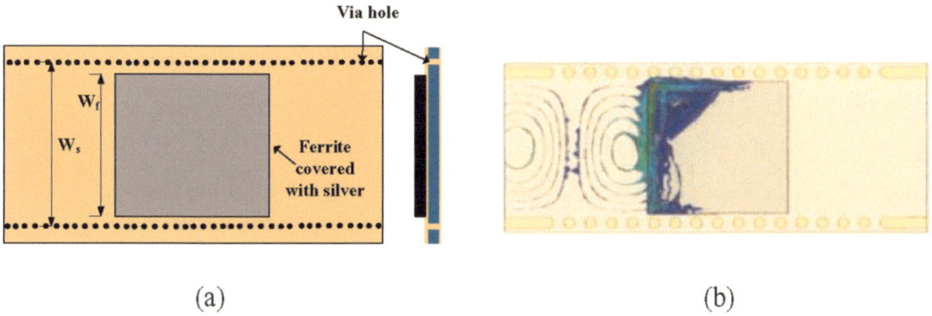

Figure 29.
(a) The proposed SIW switch (b) simulated e-fields within the switch(off-state) [37].

sidewall slots of SIW [36]. By applying external magnetic bias on the ferrite slabs, the cutoff frequency of the waveguide can be changed. The proposed switch has an IL of less than 1 dB and an isolation of 20 dB.

Figure 29a shows another magnetic control SIW switch with loading ferrite. A dc magnetic bias is applied to control the propagation capability. A Ferrite slab is placed on the surface of the SIW. The surfaces of the ferrite slab are covered with silver except for the one touching the substrate. As shown in **Figure 29b**, when the switch is off-state, e-fields present a strong transverse field displacement, causing the cutoff in TE10 mode. This way is easy to be handled by PCB technology and achieves better performance.

4.3 Mechanical switch

The following introduces a high-isolation SPDT SIW mechanical switch in V and W bands. Four via holes with the ring and cross are used in each SIW path to block the waves, as shown in **Figure 30**. A copper pad is put on top of the via holes. When a copper pad touches the SIW, via holes connect the copper pad and top layer copper. Via holes begin to block waves. When lifting the copper pad, via holes are invisible, and the switch is off-state. The proposed switch has a more than 50 dB isolation [38].

Table 3 shows the performance parameter of the switches following different technological approaches.

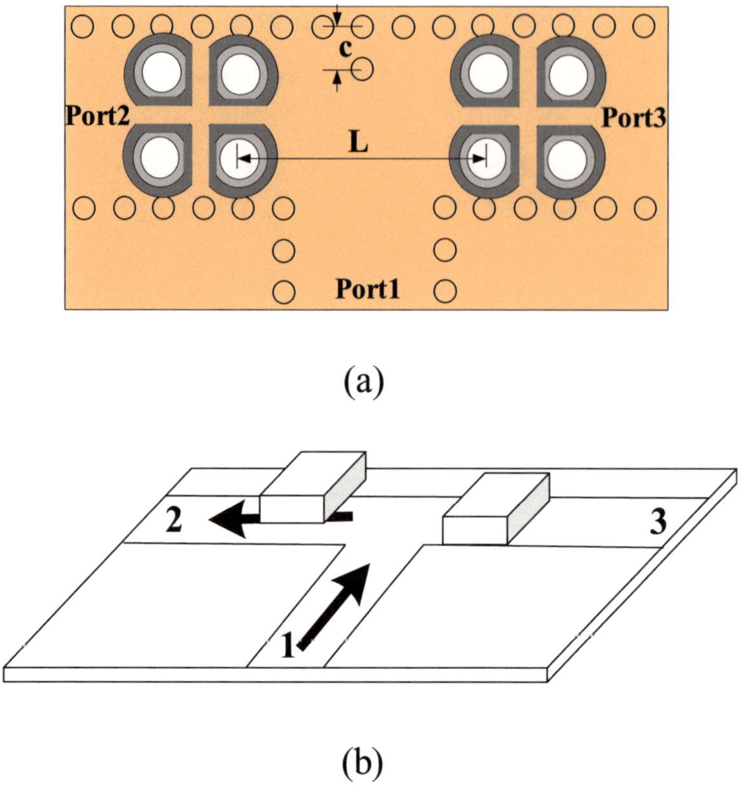

Figure 30.
(a) Mechanical SPDT switch (b) schematic of the proposed SPDT.

Device	Technology	Bandwidth (GHz)	Isolation (dB)	Insertion loss (dB)	Return loss (dB)
[32]	PIN diode / electric control	3–4.7	50	3	10
[33]	PIN diode/ electric control	8.24–10.36	15	2.55	10
[34]	PIN diode/ electric control	20–25	10	1.3	10
[35]	PIN diode/ electric control	4.6–5.3	23	1.35	15
[36]	Ferrite/magnetic control	9.5–10	20	—	10
[37]	Ferrite/magnetic control	9.5–11	27	1.6	15
[38]	Mechanical switch	50–75 75–105 105–110	50 50 50	3.5 7.5 10	10 10 10

Table 3.
Comparisons with different switches.

5. SIW phase shifter

Phase shifters are very important devices for signal phase adjustment and antenna beam steering. A SIW phase shifter has the advantages of low insertion loss, high power handling capability, and excellent immunity to electromagnetic interference

(EMI). The classical design methods of phase shifters based on SIW are mainly to control the equivalent dielectric constant of the transmission line [39–41], make self-compensating wideband shifters [42–44], and take advantage of the slow-wave effect [45, 46]. Controlling the equivalent dielectric constant of transmission is the most widely used method. There are many designs such as using air-filled slabs and loading extra metallic posts. The designed phase shifters have good performance in a narrow band, but all of them are sensitive to frequency, which has poor performance in a wide bandwidth. Self-compensating technology is used to realize wide-band phase shifters. A phase shifter that takes advantage of the slow-wave effect can get a more compact size. In addition, dynamic tunable phase shifters are also popular. The main criteria for SIW phase shifter include insertion loss, operating bandwidth, phase deviation, and power handling capabilities. Next, we will introduce several phase shifters with excellent performance.

5.1 Self-compensating wideband shifter

A phase shifter just using a delay line cannot be used to achieve the phase shift in a wide frequency range because it will cause a large phase deviation. To solve this problem, we can use the phenomenon that the delay line and the unequal-width structure to make an opposite dispersive phase shift and realize a wideband compensation phase shifter [42]. As shown in **Figure 31**, a broadband SIW self-compensating phase shifter is proposed. The differential topology reduces the frequency sensitivity. The fabricated phase shifter achieves a 90° phase shift from 25.11 to 39.75 GHz, and the phase imbalance between the two paths is within 0.2 dB and 2.5 dB, respectively [43].

Another compensation method is slotting compensation [44]. By taking advantage of the effects of the slots on the value and slope of the phase shift, slotted SIW is proposed to produce the phase shift slopes which are opposite to the microstrip delay lines to achieve a wide bandwidth phase shifter. The concept of such a technique is verified by designing −90°, −45°, 45°, and 90° phase shifter prototypes, as shown in **Figure 32**.

5.2 Dynamic tunable phase shifter

By incorporating switches into the circuits, a phase shifter can achieve multiple phase changes easily. According to the different types of switches, controllable phase shifters are divided into electrically controlled tunable phase shifter,

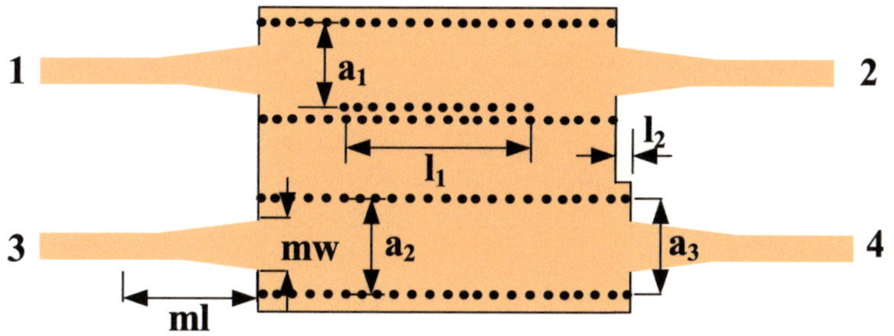

Figure 31.
Structure of the proposed SIW self-compensating phase shifter.

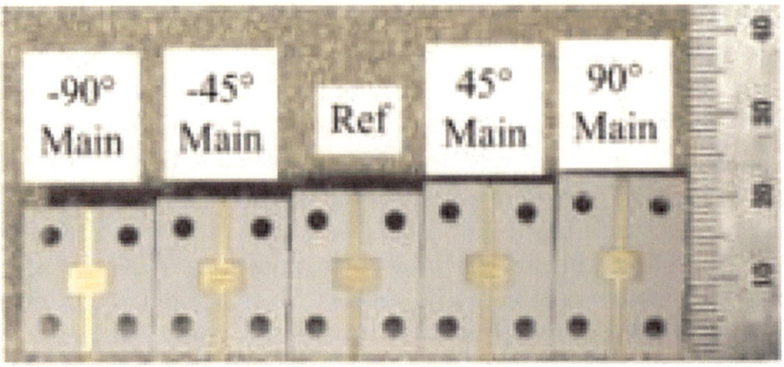

Figure 32.
Photograph of five prototypes [44].

magnetically controlled tunable phase shifter, and mechanical tunable phase shifter. Compared with the fixed phase shifter, the variable phase shifter is more flexible to meet the requirements of the system, which facilitates the evolution of phased array techniques strongly. Early SIW multiple phase shifters were designed with a multi-phase channel side by side, which increased the size [47]. At present, most of the adjustable phase shifters are single-channel structures, and the phase shift is adjusted by switching the equivalent dielectric constant of the transmission line in the channel.

A cylindrical metal post inserted in SIW is equivalent to a T-network high-pass filter [48]. When a PIN diode is buried within a SIW, the diode shows the same behavior as the metal post [49]. The difference is that diode on or off can change the value of the phase shift on the SIW. Accordingly, a new approach to designing a dynamic phase shifter is proposed. A bias circuit is placed on top of the SIW to control the diode states, as shown in **Figure 33**.

Different from diode tuning, a SIW phase shifter which can be reconfigured using liquid metal is presented [50]. A series of holes that can be filled or emptied of liquid metal is placed on the SIW. The holes with filled liquid metal can be seen as a wall to block the passage of energy. **Figure 34** illustrates the operating paths in different states. By using the different electrical lengths among the three paths, the device achieves coarse steps of phase change, from 0° up to 180°, in steps of 60°. Utilizing reactive loading (placing a single hole in each path), the phase shifter can achieve a phase shift in steps of 10°, from 0° up to 180°.

5.3 Slow wave effect

Based on the slow-wave effect, an excellent phase shifter is fabricated using a CSRR-loaded SIW, as shown in **Figure 35**. CSRR provides the slow-wave effect in the TE10 mode of SIW. The phase velocity reduction resulting from the slow-wave effect provides a phase shift [45]. By controlling the phase velocity and the physical length of the transmission line, different degrees of phase shift can be achieved. The CSRR-loaded SIW realizes a direct conversion between the fast wave and the slow wave, which makes the phase shifter integrate with other devices easily.

Table 4 shows the performance parameter of the phase shifters following different technological approaches.

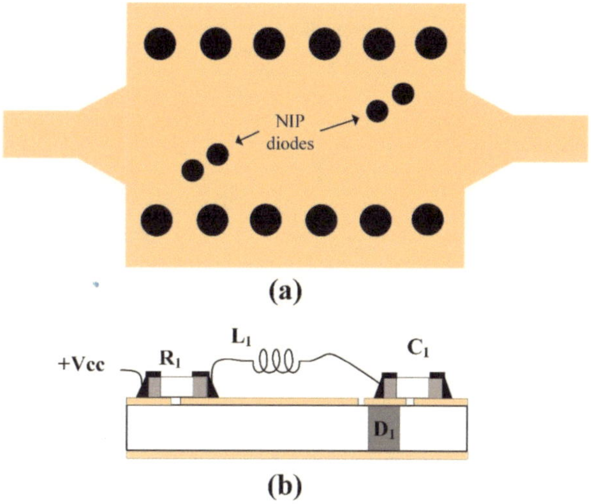

Figure 33.
Proposed SIW controllable phase shifter (a) structure. (b) Biasing circuit of the NIP diodes [49].

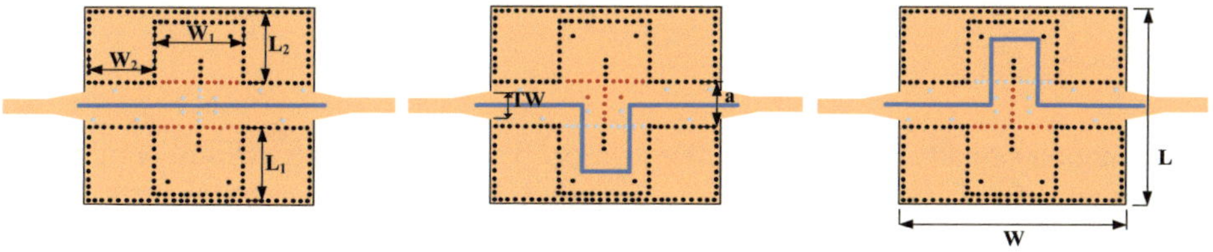

Figure 34.
The three operating states of the proposed phase shifter [50].

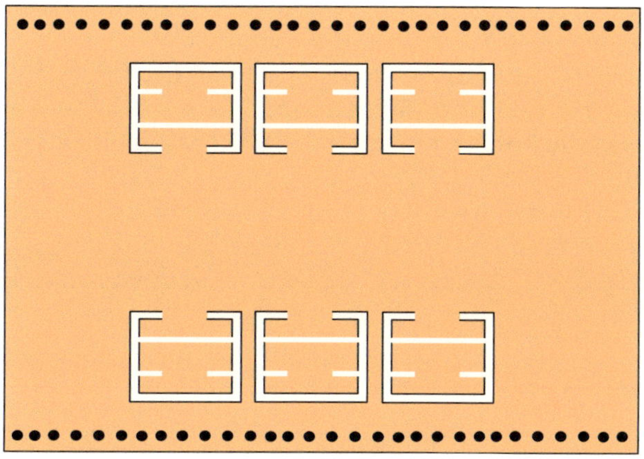

Figure 35.
Proposed a slow-wave structure.

6. SIW diplexer

The diplexer is a three-terminal device used for transmitting and receiving signals. One port is connected to the antenna and the other two are used for transmitting

Device	Technology	Bandwidth (GHz)	Phase shift (°)	Amplitude imbalance (dB)	Return loss (dB)
[39]	Equivalent dielectric constant	26.5–40	43 ± 6	0.23 ± 0.2	12.5
[40]	Equivalent dielectric constant	26.5–40	40.8 ± 2.6	—	10
[41]	Equivalent dielectric constant	20–32	45 ± 5 90 ± 5	0.1	12
[42]	Self-compensating	25.11–39.75	90.5 ± 2.5	0.2	12
[44]	Self-compensating	21.2–32.7	−90 ± 5 −45 ± 2.5 45 ± 2.5 90 ± 5	—	15
[45]	Slow-wave effect	7–10	90 ± 5	—	15
[46]	Slow-wave effect	10.5–11.5	180 ± 10	—	12
[49]	Tunable/PIN diode	10	0–45	—	10
[50]	Tunable/liquid metal	10	0–180	—	15

Table 4.
Comparisons with different phase shifters.

and receiving signals, respectively. The two processes are completed independently without any effect on each other. Thus, isolation, insertion loss, cost, and size should be paid more attention to. Diplexers can be divided into the following classes according to the shunting method of signals. Traditional diplexers mainly consist of two filters in different frequency bands and a T-junction [51–54], which increase the design complexities and the insertion loss. In recent years, there are two main ways to replace the T-junction. Firstly, multiple dual-mode resonators (MDMR) are used to replace the T-junction [55]. At the same time, good isolation is achieved through the orthogonality of the dual modes. However, the design flexibility of these diplexers is limited by the dual-mode coupling mechanisms. The second way is employing the scheme of a common dual-mode resonator (CDMR) and multiple single-mode resonators [56–59]. This method solves the problem of flexibility and achieves circuit miniaturization. Three structures are shown in **Figure 36**. In addition, there are also some other hot spots and trends, such as the miniaturization design of diplexers based on sub-mode SIW, balanced SIW diplexers, and the tunable design of SIW diplexers.

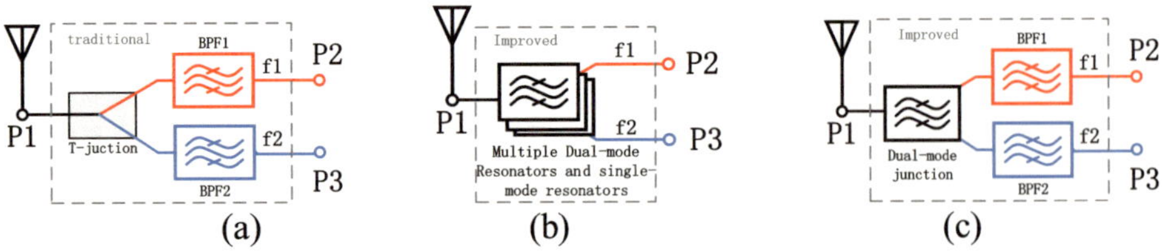

Figure 36.
(a) T-junction scheme (b) multiple dual-mode resonators scheme (c) a common dual-mode and multiple single-mode resonators scheme.

6.1 T-junction structure

A SIW diplexer with CSRRs loaded is shown in **Figure 37** [52]. A direct microstrip line insert-feeding is adopted on the top working as a T-junction. The left and right structures are two bandpass filters operating in different frequency bands.

6.2 Multiple dual-mode resonators

Figure 38 shows a six-port balanced SIW diplexer by using dual-mode resonators [55]. Because the electric field responds differently to each mode and the characteristics of differential signals, there is better isolation performance between two channels. The electric field distribution of the lower channel and the higher channel is shown in **Figure 39**. Additionally, differential input can enhance common-mode rejection and increase the stability of circuits.

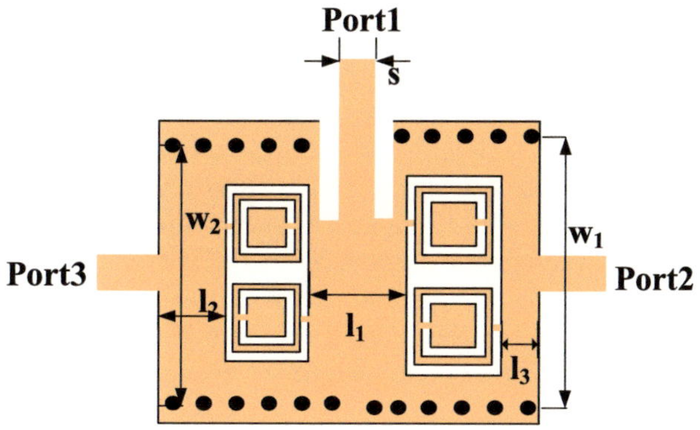

Figure 37.
Topology of the proposed SIW CSRR diplexer.

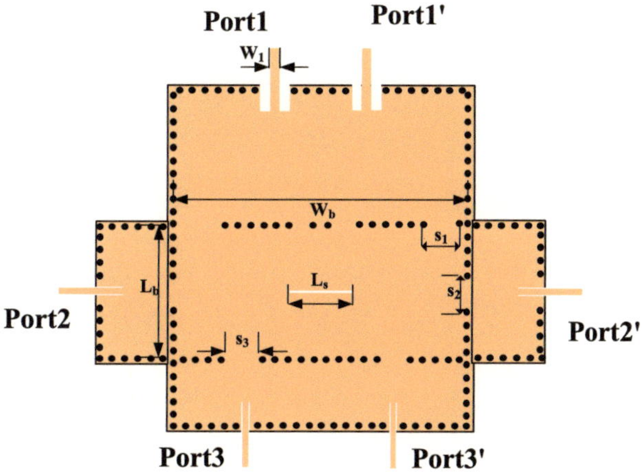

Figure 38.
Structure of the balanced diplexer.

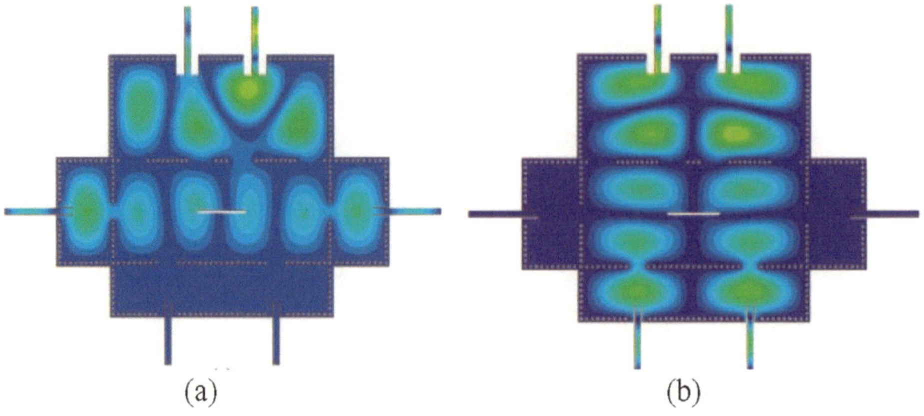

Figure 39.
Electric field distributions in (a) lower channel. (b) Higher channel [55].

6.3 A common dual-mode and multiple single-mode resonators

A wide passband SIW diplexer with a common half-mode dual-mode resonator and multiple single-mode resonators is shown in **Figure 40**. The first half-mode dual-mode resonator is used to flexibly allocate fractional bandwidths. The rest of the resonators are single-mode couplings, which almost be mapped independently. **Figure 41** shows the electric field distributions in this diplexer at 3.5 and 5.0 GHz, respectively. Different channels are dominated by different modes.

The schematic of a tunable SIW diplexer with various single-ended and balanced ports is shown in **Figure 42** [59]. Resonator A is a common dual-mode resonator, Resonator B and A form the lower channel, while Resonator C and resonator A form the upper channel. The bottom of each resonator is covered with a silver disk, which attaches to a piezo disk. Piezoelectric actuator is used to tune the operating passband. The single-end structure and balanced structure are both fabricated and have state-of-the-art performance.

Table 5 shows the performance parameter of the diplexers following different technological approaches.

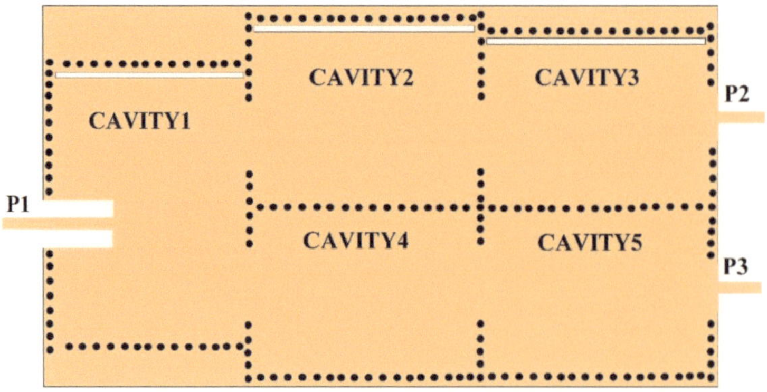

Figure 40.
Structure of the fabricated prototype [58].

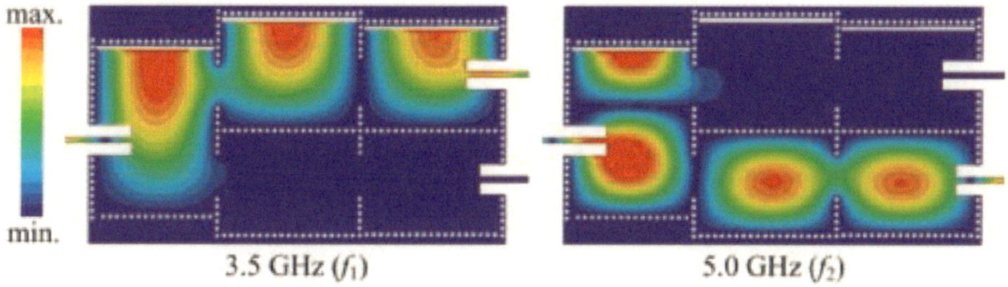

Figure 41.
Electric field distributions in the proposed SIW diplexer [58].

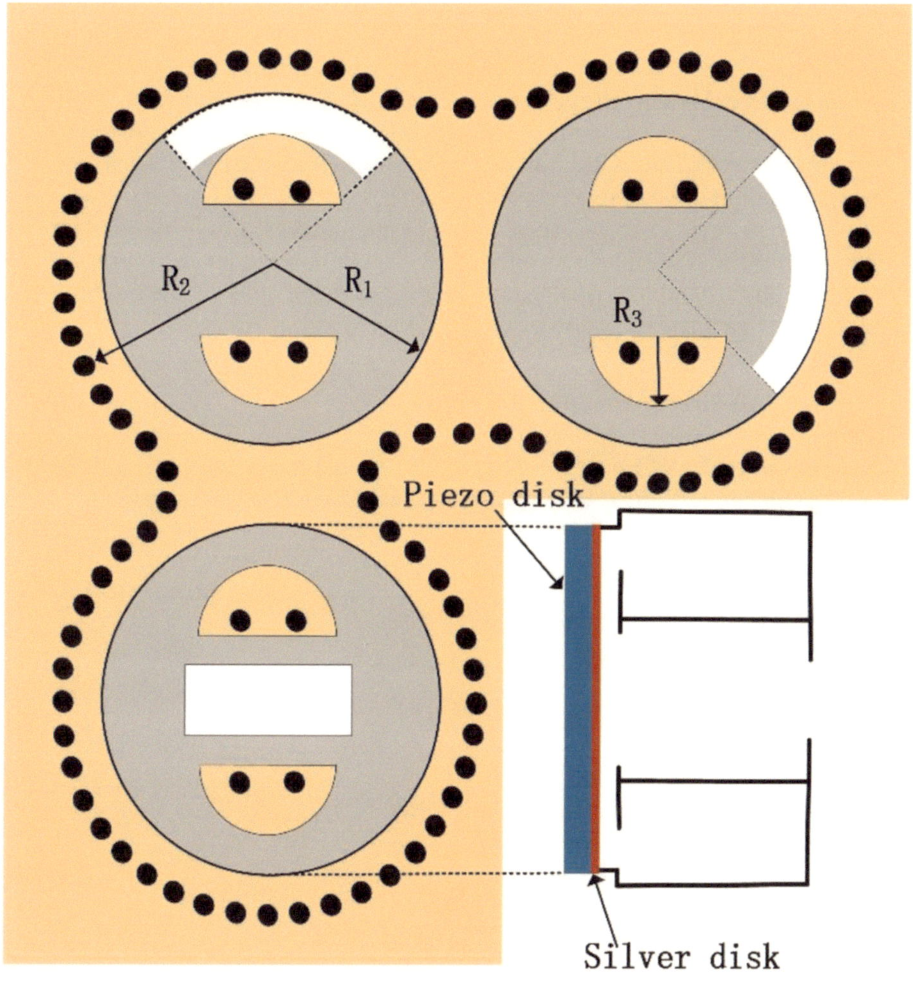

Figure 42.
The schematic of the tunable diplexer [59].

Device	Technology	Lower channel/ Upper channel (GHz)	Isolation (dB)	Size (mm × mm)	Insertion Loss (dB)
[51]	T-junction	24.31–25.66/ 25.96–27.34	50	—	1.95/2.09
[52]	T-junction	4.66/5.8	30	17.5 × 14	1.6/2.3
[54]	T-junction	24.3–25.65/ 25.95–27.35	50/40	—	2/2.5
[55]	MDMR	8.62–8.98/9.17–9.63	35	47.45 × 62.5	2.2/2.3
[56]	CDMR	8/9	40	31.13 × 31.13	2.86/3.04
[57]	CDMR	11.8–12.2/ 13.24–14.26	27	33.6 × 23.4	1.34/1.41
[58]	CDMR	3.44–3.56/5.41–5.59	10.1/21.6	87.2 × 46.4	2.77/2.55

Table 5.
Comparisons with different diplexers.

7. SIW crossover

As a microwave element, the crossover is not only able to make two signals cross-transmit and maintain high isolation, but also usually appears in the array antenna beamforming network, and is an important part of the Butler array. The use of SIW technology to design the crossover has become a research hotspot for many scholars. According to the circuit structure, the existing SIW crossovers can be divided into two categories, one is the traditional cascaded 3-dB directional coupler structure, and the other is based on the coupling theory, utilizing the orthogonality of the modes in the SIW resonators to achieve isolation of two signals.

7.1 Cascaded 3-dB directional coupler structure

The planar structure of cascading two 3-dB SIW couplers is the earliest SIW crossover. As shown in **Figure 43**, the proposed crossover has a measured return loss of lower than 13-dB, isolation of better than 20 dB, and an insertion loss of less than

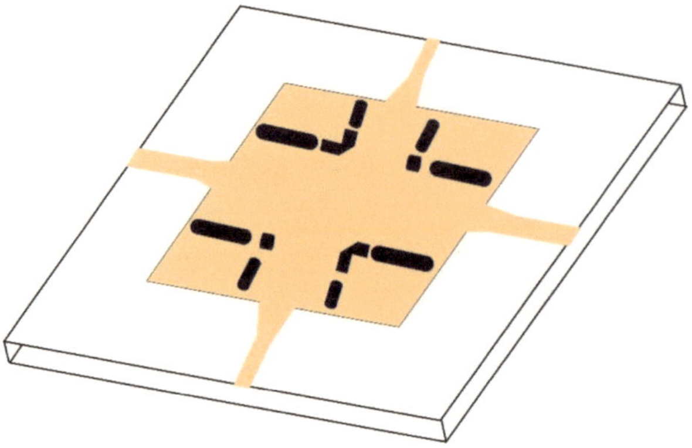

Figure 43.
The crossover at 60 GHz [60].

0.5 dB within 5% bandwidth [60]. However, cascading inevitably brings the problem of excessive circuit size. To reduce the size, a two-layer cascaded 3-dB coupler is proposed based on the two-layer folded SIW directional coupler [61]. The electric field and magnetic field are coupled through the middle slot.

7.2 SIW resonators

For the SIW resonator, on the one hand, using signal resonance to cascade multiple cavities on one channel can realize signal filtering; on the other hand, using the orthogonality of the resonant modes in the cavity, the signals in two channels can be isolated. The combination of the two can realize the dual-function integration of filtering and crossover in one device, which will reduce the complexity of the circuit and help reduce signal transmission loss.

In a general filter crossover circuit structure, according to the duality principle, each channel needs three cavities, and the design and implementation of the crossover require at least five SIW cavities, as shown in **Figure 44**.

The commonly used dual-mode SIW resonators are mostly TE201 and TE102 modes. As shown in **Figure 45**, the electric field distributions of the two modes are orthogonal to each other. The strongest electric field of one mode is the weakest of the other mode. The four ports are arranged according to the strength of the electric field, and the two signals are transmitted in the TE201 and TE102 modes respectively, which can realize the cross isolation of the signals. On this basis, a filter crossover circuit with cascading five dual-mode resonators is designed [62].

From the perspective of reducing circuit area and heat dissipation loss, the industrial field has higher and higher requirements for device performance. SIW crossovers are developing in the direction of better filtering performance, smaller size, and higher integration. Three ways to improve development are summarized below.

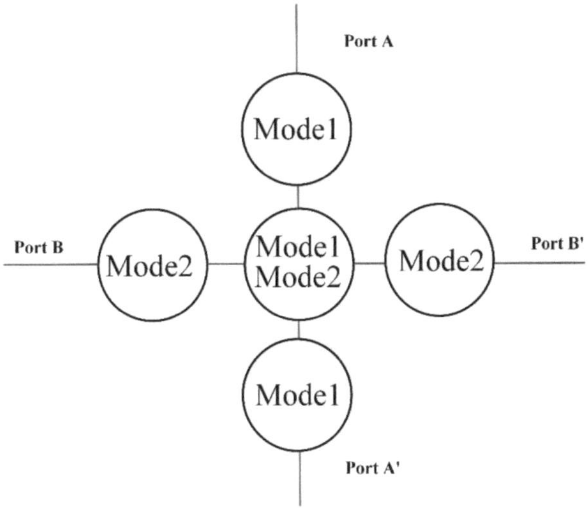

Figure 44.
Traditional filter crossover.

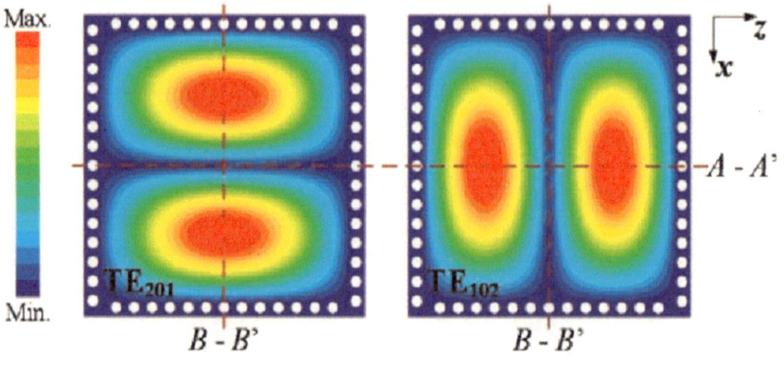

Figure 45.
Distributed electric fields of TE102 and TE201 in rectangular SIW cavity.

7.3 Filtering crossover

In the traditional filter crossover circuit with five cavities, to further reduce the circuit size, the four peripheral dual-mode cavities are changed to single-mode resonators whose fundamental mode is TE101 [63, 64], and the cavity arrangement is different, but the signal filtering selectivity of the crossover is enhanced in [64]. By increasing the number of cavities, and flexibly arranging single-mode and dual-mode resonator cavity arrangements, a variety of filter crossovers with different center frequencies and bandwidths are realized [65]. The circuit only retains one dual-mode resonator, and four CPW half-wave resonators are cascaded at the four ports, which further reduces the circuit size [66]. In addition, square SIW cavities with TE102 and TE201 modes are used as building blocks, and multi-channel filtering crossover is designed and fabricated by rationally arranging coupling ports and feed ports [67].

Figure 46 shows the electric field distribution of a filter crossover circuit with a single resonator. The circuit is based on a four-mode resonant SIW rectangular reso-

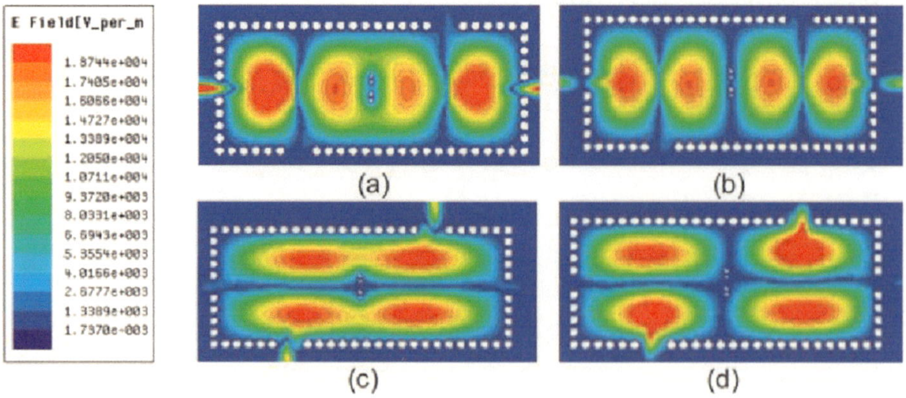

Figure 46.
Electric field distribution in the crossover: (a) TE103 mode, (b) TE104 mode, (c) TE201 mode, (d) TE202 mode [68].

nant cavity, and the middle through holes and slots are used to adjust the electric field distribution. Compared with the above-mentioned circuits with multiple resonant cavities, on the one hand, the structure only uses a single resonant cavity, which further reduces the size, and on the other hand, the operating frequencies of the two channels in the above structure are the same, but the center frequencies of the two channels in this structure are different, which realizes the dual-frequency crossover of the two-channel signals and expands the application range.

In [69], the circuit structure is still in the traditional cross form, but the SIW evanescent-mode (EVA) resonator is applied. The coupling coefficient between the resonators is controlled by diodes, which makes the coupling coefficient of the adjacent cavities 0, and the diagonal cavities are not 0, thus completing the isolation of two signals. Piezoelectric actuators are utilized to form a tunable filtering channel, and the application range of the crossover is expanded.

7.4 Crossover with balanced structure

The above-mentioned crossover circuits are all used for single-ended circuits, but considering the reduction of signal noise interference, the crossover circuits applied to balanced circuits are also studied. The multi-layer circuit topology is the main structure.

Based on the traditional filter crossover circuit, five SIW cavities are cascaded as shown in **Figure 47**. The CM is suppressed in this structure. Under differential mode (DM), the crossover part is realized by orthogonal modes TE102 and TE201.

Another balanced structure is shown in **Figure 48**. The 1–3 channels and the 2–4 channels are arranged in different layers, and the signals are transmitted through the middle slots. TE202 and TE204 in a single rectangular cavity are orthogonal to each other to complete the cross isolation of the two signals.

Table 6 shows the performance parameter of the crossovers following different technological approaches.

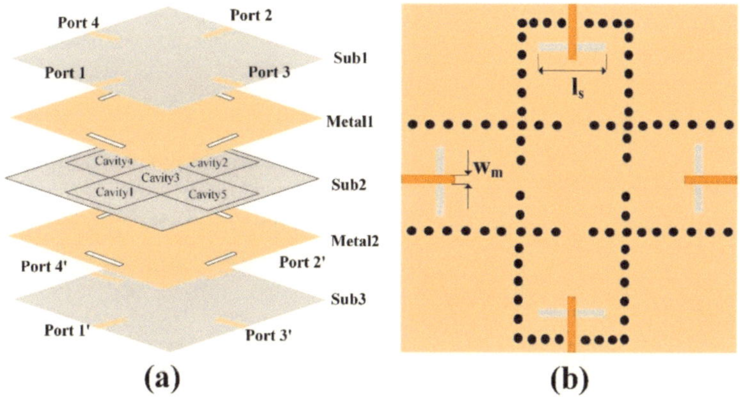

Figure 47.
The crossover circuit layout [70]. (a) 3D figure; (b) Planar graph.

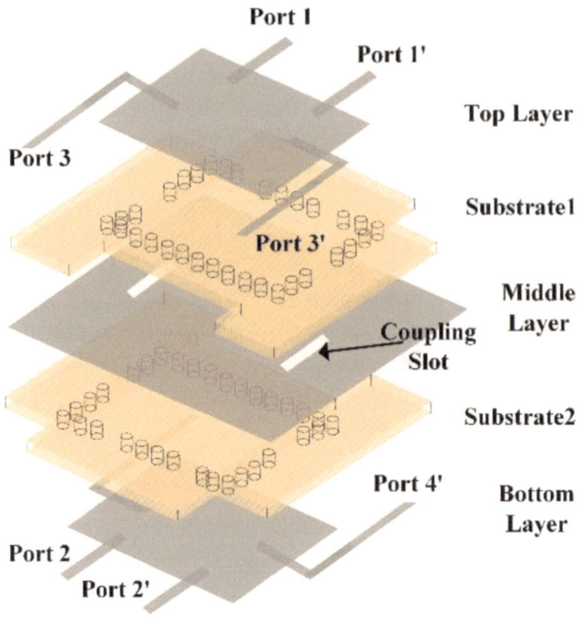

Figure 48.
3D diagram of crossover [71].

Device	Technology	Bandwidth (GHz)	Isolation (dB)	Insertion loss (dB)	Return loss (dB)
[60]	3-dB Directional coupler	5	20	0.5	13
[65]	Filtering crossover	11.57–12.43	21.8	1.61	18.5
[67]	Filtering crossover	8.83	23	2.29	20
[68]	Filtering crossover	7.11/8.77	14.65	2.7/2.57	13.1
[70]	Balanced structure	10.2	30	3.2	—

Table 6.
Comparisons with different crossovers.

8. Conclusions

As an easy-to-integrate planar circuit, the SIW technology lays the foundation for the design and implementation of high-performance, planar, low-cost, and easy-to-integrate microwave passive circuits. This chapter describes the novel circuits based on substrate-integrated waveguide, including filter, power divider, switch, phase shifter, duplexer, crossover, and so on.

Author details

Zhongmao Li[1,2*], Mengjie Qin[1,2], Pengzhan Liu[1,2] and Xin Qiu[1,2]

1 Institute of Microelectronics, Chinese Academy of Sciences, Beijing, China

2 School of Integrated Circuits, University of Chinese Academy of Sciences, Beijing, China

*Address all correspondence to: lizhongmao@ime.ac.cn

IntechOpen

© 2022 The Author(s). Licensee IntechOpen. This chapter is distributed under the terms of the Creative Commons Attribution License (http://creativecommons.org/licenses/by/3.0), which permits unrestricted use, distribution, and reproduction in any medium, provided the original work is properly cited.

References

[1] Hong W. Research Advances in SIW, HMSIW and FHMSIW. In: 2008 China-Japan Joint Microwave Conference. 2008. pp. 357–358

[2] Hirokawa J, Ando M. Single-layer feed waveguide consisting of posts for plane TEM wave excitation in parallel plates. IEEE Transactions on Antennas and Propagation. 1998;**46**(5):625-630

[3] Wu K, Deslandes D, Cassivi Y. The substrate integrated circuits - a new concept for high-frequency electronics and optoelectronics. In: 6th International Conference on Telecommunications in Modern Satellite, Cable and Broadcasting Service. 2003

[4] Deslandes D, Wu K. Accurate modeling, wave mechanisms, and design considerations of a substrate integrated waveguide. IEEE Transactions on Microwave Theory and Techniques. 2006;**54**(6):2516-2526

[5] Hao ZC, Hong W, Chen XP, Chen JX, Wu K, Cui TJ. Multilayered substrate integrated waveguide (MSIW) elliptic filter. IEEE Microwave and Wireless Components Letters. 2005;**15**(2):95-97

[6] Deslandes D, Wu K. Single-substrate integration technique of planar circuits and waveguide filters. IEEE Transactions on Microwave Theory and Techniques. 2003;**51**(2):593-596

[7] Zhu F, Hong W, Chen J, Wu K. Wide stopband substrate integrated waveguide filter using corner cavities. Electronics Letters. 2013;**49**(1):50-51

[8] Grigoropoulos N, Sanz-Izquierdo B, Young PR. Substrate integrated folded waveguides (SIFW) and filters. IEEE Microwave and Wireless Components Letters. 2005;**15**(12):829-831

[9] Chien H, Shen T, Hung T, Wu R. Design of a vertically stacked substrate integrated folded-waveguide resonator filter in LTCC. In: 2007 Asia-Pacific Microwave Conference. 2007. pp. 1–4

[10] Wenquan Che, Cuixia Li, Russer P, Chow YL. Propagation and band broadening effect of planar integrated ridged waveguide in multilayer dielectric substrates. In: 2008 IEEE MTT-S International Microwave Symposium Digest. 2008. pp. 217–220

[11] Yang YN, Li GH, Sun L, Yang W, Yang XX. Design of compact Bandpass filters using sixteenth mode and thirty-second mode SIW cavities. Progress in Electromagnetics Research Letters. 2018; **75**:61-66

[12] Zheng Y, Zhu Y, Wang Z, Dong Y. Compact, wide stopband, shielded hybrid filter based on quarter-mode substrate integrated waveguide and microstrip line resonators. IEEE Microwave and Wireless Components Letters. 2020;**31**(3):245-248

[13] Hong W, Liu B, Wang Y, Lai Q, Tang H, Yin XX et al. Half mode substrate integrated waveguide: A new guided wave structure for microwave and millimeter wave application. In: 2006 Joint 31st International Conference on Infrared Millimeter Waves and 14th International Conference on Teraherz Electronics. 2006. pp. 219–219

[14] Chen F, Song K, Hu B, Fan Y. Compact dual-band bandpass filter using HMSIW resonator and slot perturbation. IEEE Microwave and Wireless Components Letters. 2014;**24**(10): 686-688

[15] Zhenyu Zhang, Ning Yang, K. Wu. 5-GHz bandpass filter demonstration using quarter-mode substrate integrated waveguide cavity for wireless systems. In: 2009 IEEE Radio and Wireless Symposium. 2009. pp. 95–98

[16] Moscato S, Tomassoni C, Bozzi M, Perregrini L. Quarter-mode cavity filters in substrate integrated waveguide technology. IEEE Transactions on Microwave Theory and Techniques. 2016;**64**(8):2538-2547

[17] Wang X, Zhu X-W, Jiang ZH, Hao Z-C, Wu Y-W, Hong W. Analysis of eighth-mode substrate-integrated waveguide cavity and flexible filter design. IEEE Transactions on Microwave Theory and Techniques. 2019;**67**(7): 2701-2712

[18] Azad A, Mohan A. A compact sixteenth-mode substrate integrated waveguide bandpass filter with improved out-of-band performance. Microwave and Optical Technology Letters. 2017;**59**(7):1728-1733

[19] Zhang S, She J, Tong M, Hong J. Mode selective bandpass filter with high selectivity based on thirty-second-mode circular SIW resonator. Microwave and Optical Technology Letters. 2021;**63**(7): 1820-1825

[20] Pendry JB, Holden AJ, Robbins DJ, Stewart WJ. Magnetism from conductors and enhanced nonlinear phenomena. IEEE Transactions on Microwave Theory and Techniques. 1999;**47**(11):2075-2084

[21] Huang Y, Peng Y, Zhou Y, Jin H, Leng S, Wang G. Size-reduced dual-band HMSIW cavity filters loaded with double-sided SICSRRs. Electronics Letters. 2017 11;**53**(10):689-690

[22] Senior DE, Cheng X, Yoon YK. Dual-band filters using complementary split-ring resonator and capacitive loaded half-mode substrate-integrated-waveguide. In: Proceedings of the 2012 IEEE International Symposium on Antennas and Propagation. 2012. pp. 1–2

[23] Mira F, Mateu J, Collado C. Mechanical tuning of substrate integrated waveguide resonators. IEEE Microwave and Wireless Components Letters. 2012;**22**(9):447-449

[24] Karpuz C, Unuk GBF, PÖ Özdemir. Design of dual-mode tunable filter surrounded by an electrical wall to obtain shielding effect. In: 2017 47th European Microwave Conference (EuMC). 2017. p. 25–28

[25] Moulay A, Djerafi T. Wilkinson power divider with fixed width substrate-integrated waveguide line and a distributed isolation resistance. IEEE Microwave and Wireless Components Letters. 2018;**28**(2):114-116

[26] Song K, Xia F, Zhou Y, Guo S, Fan Y. Microstrip/slotline-coupling substrate integrated waveguide power divider with high output isolation. IEEE Microwave and Wireless Components Letters. 2019;**29**(2):95-97

[27] Pradhan NC, Subramanian KS, Barik RK, Cheng QS. Design of compact substrate integrated waveguide based triple- and quad-band power dividers. IEEE Microwave and Wireless Components Letters. 2021;**31**(4):365-368

[28] Chi P, Chen Y, Yang T. Single-layer dual-band balanced substrate- integrated waveguide filtering power divider for 5G millimeter-wave applications. IEEE Microwave and Wireless Components Letters. 2020;**30**(6):585-588

[29] Zheng J, Zhang G, Zhou X, Xie C, Zhang Y, Wang X, et al. Dual-band filtering power divider based on CSRRs

loaded SIW cvity. IEEE Transactions on Circuits and Systems II: Express Briefs. 2022;**69**(2):394-398

[30] Liu B-G, Lyu Y-P, Zhu L, Cheng C-H. Compact square substrate integrated waveguide filtering power divider with wideband isolation. IEEE Microwave and Wireless Components Letters. 2021; **31**(2):109-112

[31] Pan BC, Yu P, Liao Z, Zhu F, Luo GQ. A compact filtering power divider based on spoof surface plasmon polaritons and substrate integrated waveguide. IEEE Microwave and Wireless Components Letters. 2022; **32**(2):101-104

[32] Xu RF, Izquierdo BS, Young PR. Switchable substrate integrated waveguide. IEEE Microwave and Wireless Components Letters. 2011; **21**(4):194-196

[33] Lim I, Lim S. Substrate-Integrated-Waveguide (SIW) single-pole-double-throw (SPDT) switch for X-band applications. IEEE Microwave and Wireless Components Letters. 2014; **24**(8):536-538

[34] Numan AB, Frigon J-F, Laurin J-J. Single-pole single-throw switch for substrate-integrated waveguide. IEEE Microwave and Wireless Components Letters. 2018;**28**(3):221-223

[35] Chen H, Che W, Cao Y, Feng W, Sarabandi K. Function-reconfigurable between SPDT switch and power divider based on switchable HMSIW unit. IEEE Microwave and Wireless Components Letters. 2017;**27**(3):275-277

[36] Ghiotto A, Adhikari S, Wu K. Ferrite-loaded substrate integrated waveguide switch. IEEE Microwave and Wireless Components Letters. 2012; **22**(3):120-122

[37] Huang QD, Cheng YJ, Li JL-W. Surface-loaded ferrite substrate integrated waveguide switch. IEEE Microwave and Wireless Components Letters. 2015;**25**(4):232-234

[38] Wei S-C, Yang C-H, Chen Y-C, Chen T-A, Chang C-Y. V - and W -Band substrate integrated waveguide (SIW) mechanical switch. IEEE Transactions on Microwave Theory and Techniques. 2018;**66**(6):3090-3098

[39] Parment F, Ghiotto A, Vuong T-P, Duchamp J-M, Wu K. Double dielectric slab-loaded air-filled SIW phase shifters for high-performance millimeter-wave integration. IEEE Transactions on Microwave Theory and Techniques. 2016;**64**(9):2833-2842

[40] Boudreau I, Wu K, Deslandes D. Broadband phase shifter using air holes in substrate integrated waveguide. In: 2011 IEEE MTT-S International Microwave Symposium. 2011. pp. 1–1

[41] Djerafi T, Wu K, Tatu S. Substrate-integrated waveguide phase shifter with rod-loaded artificial dielectric slab. Electronics Letters. 2015;**51**(9):707-708

[42] Cheng YJ, Wu K, Hong W. Substrate integrated waveguide (SIW) broadband compensating phase shifter. In: 2009 IEEE MTT-S International Microwave Symposium Digest. 2009. pp. 45–848

[43] Cheng YJ, Hong W, Wu K. Broadband self-compensating phase shifter combining delay line and equal-length unequal-width phaser. IEEE Transactions on Microwave Theory and Techniques. 2010;**58**(1):203-210

[44] Zhang W, Shen Z, Xu K, Shi J. A compact wideband phase shifter using slotted substrate integrated waveguide. IEEE Microwave and Wireless

Components Letters. 2019;**29**(12): 767-770

[45] Liu S, Xu F. Novel substrate-integrated waveguide phase shifter and its application to six-port junction. IEEE Transactions on Microwave Theory and Techniques. 2019;**67**(10): 4167-4174

[46] Bing Liu W, Zhang Cheng Hao K. Wu. Substrate integrated waveguide 180-degree narrow-wall directional coupler. In: 2005 Asia-Pacific Microwave Conference Proceedings. 2005. p. 3

[47] Yang T, Ettorre M, Sauleau R. Novel phase shifter design based on substrate-integrated-waveguide technology. IEEE Microwave and Wireless Components Letters. 2012;**22**(10):518-520

[48] Sellal K, Talbi L, Denidni T, Lebel J. Design and implementation of a substrate integrated waveguide phase shifter. IET Microwaves Antennas & Propagation. 2008;**2**(2):194-199

[49] Sellal K, Talbi L, Nedil M. Design and implementation of a controllable phase shifter using substrate integrated waveguide. IET Microwaves Antennas & Propagation. 2012;**6**(9): 1090-1094

[50] Alkaraki S, Borja AL, Kelly JR, Mittra R, Gao Y. Reconfigurable liquid metal-based SIW phase shifter. IEEE Transactions on Microwave Theory and Techniques. 2022;**70**(1):323-333

[51] Tang HJ, Hong W, Chen J, Luo GQ, Wu K. Development of millimeter-wave planar diplexers based on complementary characters of dual-mode substrate integrated waveguide filters with circular and elliptic cavities. IEEE Transactions on Microwave Theory and Techniques. 2007;**55**(4):776-782

[52] Dong Y, Itoh T. Substrate integrated waveguide loaded by complementary split-ring resonators for miniaturized diplexer design. IEEE Microwave and Wireless Components Letters. 2011; **21**(1):10-12

[53] Meng-Hsuan Chen, Hsiang-Sheng Cheng, Po-Jung Chou, Chi-Yang Chang. W-band T-Junction and bifurcated substrate integrated waveguide diplexers. In: 2015 International Workshop on Electromagnetics: Applications and Student Innovation Competition (iWEM). 2015. 1–2

[54] Chu P, Hong W, Tuo M, Zheng K-L, Yang W-W, Xu F, et al. Dual-mode substrate integrated waveguide filter with flexible response. IEEE Transactions on Microwave Theory and Techniques. 2017;**65**(3):824-830

[55] Song K, Yao J, Chen Y, Zhou Y, Zhu Y, Fan Y. Balanced diplexer based on substrate integrated waveguide dual-mode resonator. IEEE Transactions on Microwave Theory and Techniques. 2020;**68**(12):5279-5287

[56] Cheng F, Lin X, Song K, Jiang Y, Fan Y. Compact diplexer with high isolation using the dual-mode substrate integrated waveguide resonator. IEEE Microwave and Wireless Components Letters. 2013;**23**(9):459-461

[57] Zhou K, Zhou C-X, Wu W. Compact SIW diplexer with flexibly allocated bandwidths using common dual-mode cavities. IEEE Microwave and Wireless Components Letters. 2018;**28**(4):317-319

[58] Zhou K, Wu K. Miniaturized diplexers with large frequency ratios based on common half-mode dual-mode siw junction-cavities. IEEE Transactions on Microwave Theory and Techniques. 2021;**69**(12):5343-5350

[59] Hagag MF, Abu Khater M, Hickle MD, Peroulis D. Tunable SIW cavity-based dual-mode diplexers with various single-ended and balanced ports. IEEE Transactions on Microwave Theory and Techniques. 2018;**66**(3):1238-1248

[60] Djerafi T, Wu K. 60 GHz substrate integrated waveguide crossover structure. In: 2009 European Microwave Conference (EuMC). 2009. pp. 1014–1017

[61] Guntupalli AB, Djerafi T, Wu K. Ultra-compact millimeter-wave substrate integrated waveguide crossover structure utilizing simultaneous electric and magnetic coupling. In: 2012 IEEE/MTT-S International Microwave Symposium Digest. 2012. pp. 1–3

[62] Han S, Zhou K, Zhang J, Zhou C, Wu W. Novel substrate integrated waveguide filtering crossover using orthogonal degenerate modes. IEEE Microwave and Wireless Components Letters. 2017;**27**(9):803-805

[63] Zhou Y, Zhou K, Zhang J, Zhou C, Wu W. Miniaturized substrate integrated waveguide filtering crossover. In: 2017 IEEE Electrical Design of Advanced Packaging and Systems Symposium (EDAPS). 2017. pp. 1–3

[64] Zhou Y, Zhou K, Zhang J, Wu W. Substrate-integrated waveguide filtering crossovers with improved selectivity. International Journal of RF and Microwave Computer-Aided Engineering. 2020;**30**:e22067

[65] Zhou K, Wu K. Wide-stopband substrate-integrated waveguide filtering crossovers with flexibly allocated channel frequencies and bandwidths. IEEE Transactions on Microwave Theory and Techniques. 2021;**69**(7): 3264-3274

[66] Zhou K, Wu K. Compact substrate-integrated waveguide filtering crossover by embedding CPW quarter-wavelength resonators. In: 2020 IEEE/MTT-S International Microwave Symposium (IMS). 2020. pp. 916–919

[67] Zhan W-L, Xu J-X, Zhao X-L, Hu B-J, Zhang XY. Substrate integrated waveguide multi-channel filtering crossover with extended channel number and controllable frequencies. IEEE Transactions on Circuits and Systems II: Express Briefs. 2020;**67**(12): 2858-2862

[68] Qu L, Zhang Y, Liu J, Bo X, Jing H, Fan Y. Design of filtering crossover and diplexer on SIW quadruple-mode resonators in a single cavity. IEEE Access. 2020;**8**:176789-176796

[69] Lai J, Yang T, Chi P-L, Xu R. A Novel 1.7–2.85-GHz filtering crossover with independently tuned channel passbands and reconfigurable filtering power-dividing function. IEEE Transactions on Microwave Theory and Techniques. 2021;**69**(5):2458-2469

[70] Sun L, Deng H-W, Xue Y-F, Zhu J-M, Xing S-B. Compact-balanced BPF and filtering crossover with intrinsic common-mode suppression using single-layered SIW cavity. IEEE Microwave and Wireless Components Letters. 2020; **30**(2):144-147

[71] Zhou X, Zhang G, Zheng M, Li M. Millimeter-wave balanced filtering crossover using multi-layered SIW cavity. In: 2021 Cross Strait Radio Science and Wireless Technology Conference (CSRSWTC). 2021. pp. 345–347

Chapter 2

Challenges and Perspectives for SIW Hybrid Structures Combining Nanowires and Porous Templates

Vivien Van Kerckhoven, Luc Piraux and Isabelle Huynen

Abstract

The chapter presents a review of properties and applications of a particular category of Substrate Integrated Waveguide (SIW) named NWSIW, for NanoWire-based Substrate Integrated Waveguide. The NWSIW topology combines metallic nanowires embedded in a porous template in order to form planar compact integrated waveguides; nanowires selectively grown in the template are used for building the walls of the waveguide, but also allow to achieve microwave functionalities such as filtering, isolation. Through the chapter the comparison made with classical waveguides including SIW will lead to a discussion on future perspectives and possible improvements of this NWSIW topology. Their performances will be explained and illustrated with regards to current state-of-the art, and based on results obtained at UCLouvain.

Keywords: SIW, porous template, metallic nanowires, microwaves, isolator, filter, design, fabrication

1. Introduction

High-quality transmission lines become mandatory to fulfill the requirements of the nowadays applications. Indeed, the constraints applied to microwave and millimeter-wave circuits are continuously reviewed and increased: low cost, low consumption, high density, high operational frequency, ... At frequencies higher than 30 GHz, interferences, radiation and material losses prevent the use of conventional microstrip and coplanar transmission lines. Within this context, the substrate integrated waveguide (SIW) topology, introduced in [1], represents a promising perspective. This technology uses two rows of vertical metallic wires or metallized via holes inserted throughout the substrate entire height to form a shielded line of rectangular section, reducing subsequently radiation and conductor losses. The presence of walls also prevents spurious crosstalk interferences between devices in a same circuit [2–4].

The two virtual walls have to be carefully placed to allow the appearance of the same propagation modes as in a macroscopic rectangular waveguide. It is possible to integrate various components in a same substrate, including passive or active devices, as well as antennas. Different variations of the concept emerged for several functionalities (filter, coupler, antenna power supply, ...) with applications up to 180 GHz, and in different types of substrates (PCB, paper, polymers or alumina) [5, 6].

Over the last decades also, the interactions between nanowires embedded in porous templates and microwave signals have been intensively studied and exploited to create planar monolithic devices. Both the permeability and permittivity of the template filled with nanowires can be modulated in order to create various microwave and millimeter-wave components. The advantages of microwave devices based on nanowires compared to classical components are manifold: wide range of operating frequencies, temperature stability, monolithic integration into a single substrate and possible miniaturization. Moreover, compared to classical ferrites, ferromagnetic nanowire arrays display higher saturation magnetization and ferromagnetic resonance (FMR) frequencies as well as high operation frequencies at remanence due to their large aspect ratio.

Thanks to those properties, noise suppressors [7], absorbers [8, 9], inductors [10], or filters based on electromagnetic bandgap effect [11] have been developed, together with non-reciprocal devices, such as phase shifters [12], isolators [13] and circulators [14]. More recently slow-wave transmission lines were designed exploiting the high permittivity of metallic nanowire arrays [15], while a filter was designed using the double ferromagnetic resonance effect in magnetic nanowires [16]. The reported devices use microstrip or coplanar waveguide topologies combined with different ferromagnetic materials for the nanowires.

Given the respective advantages of nanowires and SIW [17–19], using nanowires to conceive Substrate Integrated Waveguide (SIW) devices is an interesting alternative. The basic idea underlying this concept is to use metallic nanowire arrays electrodeposited into a porous template to form the waveguide walls. With two copper layers deposited onto the template's both faces, a simple rectangular waveguide is created, denoted NWSIW for nanowire-based substrate integrated waveguide. To achieve different microwave functionalities, various nanowire arrays can be added inside the cavity of the NWSIW, combining different heights, shapes and materials (magnetic or not).

Concepts presented in this chapter are widely inspired by the work of Van Kerckhoven et al. [17–22]. The chapter is organized as follows; Section 2 introduces the Nanowired-based Substrate Integrated Waveguide (NWSIW) topology and compares it with the classical SIW and metallic rectangular waveguide (MRW) geometries. Section 3 details the design features of the NWSIW, while Section 4 discusses fabrication techniques. Section 5 presents some experimental realizations of microwave devices based on NWSIW topology, while Section 6 discusses challenges and perspectives of this new technology.

2. Geometries in presence

Figure 1 shows the transformation of a classical metallic rectangular waveguide (MRW) into an SIW, next into a NWSIW.

The MRW is formed by the assembly of four metallic plates acting as perfect electric conductors and forming a hollow metallic pipe of rectangular section for the guidance of electromagnetic waves.

The SIW has a fully planar thin topology since it bases on planar dielectric substrates commonly used for the fabrication of RF and microwave devices and circuits. The top and bottom faces of the substrate are covered by a metallic clad typically 17 or 35 μm-thick, as available for commercial copper clad laminates (CCL). These layers form the top and bottom walls of the waveguide. Lateral walls are obtained by

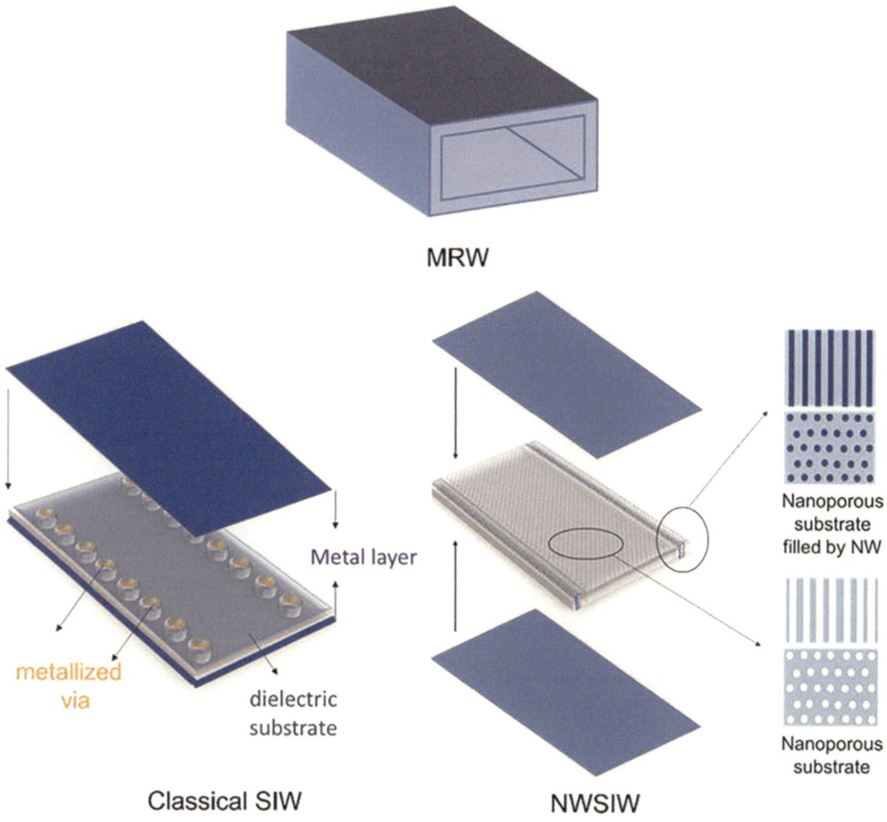

Figure 1.
From metallic rectangular waveguide (MRW) to Nano-wire substrate integrated waveguide (NWSIW). In SIW lateral metal plates of MRW are replaced by two rows of metallized vias. In NWSIW rows of nanowires (NW) electrodeposited in nanoporous substrate form continuous metallic walls. Vertical arrows indicate the positioning of covering metal plates (blue).

mechanically drilling two rows of holes that are next filled by metallic vias; periodicity and diameter of the vias must be adjusted in order to achieve a similar metallic shielding level as in an MRW.

In a NWSIW the difference with respect to SIW is the nanoporous substrate used for its design and fabrication. Instead of drilling vias in a CCL substrate, metallic nanowires (NW) are grown by electrodeposition in dedicated row areas in the nanoporous substrate in order to form a virtual metallic lateral shielding. The bottom side of the waveguide is formed by the thin metallic layer electroplated as ground electrode for the electrodeposition of NW, while top side is achieved by another electroplating realized after the growth of NW is completed.

3. Design rules and constraints for NWSIW

3.1 Alumina porous template

The template considered for the NWSIW is nanoporous alumina wherein metallic nanowires can grow. Commercial solutions are based on anodic alumina oxide (AAO);

Figure 2.
Schematic top view (left) and side view (right) of AAO template. T is the thickness of the template, D is the diameter of the pores.

Pore diameter range D	30 nm – 1 um
Type and conductivity of NW	Cu - 5.65 x 10^7 S/m
Porosity range (P)	4% - 60%
Dielectric constant bulk alumina (ε_r)	9.8
Relative permeability bulk alumina (μ_r)	1
Loss tangent factor bulk alumina (tanδ)	0.015
Thickness T	50 μm or 100 μm
Operational frequency range	1 GHz – 50 GHz

Table 1.
Characteristics of AAO porous template.

the oxide is obtained by anodization of aluminum. The result is an array of vertical cylindrical pores arranged as a hexagonal pattern, as illustrated in **Figure 2**. The diameter of the pores is fixed by the process and typically ranges between 25 nanometers and 0.5 μm. Another characteristic parameter is the porosity P of the template, defined as the total surface occupied by pores in a given area reported to the surface of this area. The available thicknesses of the AAO template and its electrical parameters are also reported in **Table 1**.

3.2 Building of NWSIW walls

Contrarily to classical a SIW where walls are created by the insertion of metallic vias, the NWSIW architecture exploits the nanoporous nature of AAO to grow arrays of metallic nanowires (MNW) inside pores. Instead of bulk metal for classical RW having bulk conductivity σ, each wall of the NWSIW formed by the nanowire array has a total equivalent conductivity reduced by a factor P [20]. As a consequence, the skin depth δ_{NW} of the nanowire array expresses as

$$\delta_{NW} = \frac{2}{\omega\mu P\sigma} \tag{1}$$

The extinction thickness ET is here defined as the thickness of the nanowire array forming the MNW wall that ensures that 99.9% of the electromagnetic field is blocked/attenuated by the MNW wall:

$$0.001 = e^{-ET/\delta_{MNW}} \rightarrow \mathrm{ET} = 7\ \delta_{\mathrm{MNW}} \tag{2}$$

Figure 3 shows the extinction thickness ET as a function of porosity, for different values of operating frequency, from DC up to 50 GHz, and for Copper (Cu) nanowires of conductivity given in **Table 1**. It is concluded that a thickness of 10 μm is sufficient to satisfy condition (2) for most values of porosity higher than 3% and frequency above 1 GHz.

Practically two rows of MNW are grown in the AAO template, having a 10 μm width and a length equal to the desired length of the NWSIW. The nanoporous AAO template imposes a close packing of the electrodeposited MNW array from which an efficient shield is obtained. This is different and easier than designing the diameter and spacing of metallic vias drilled in the dielectric substrate of classical SIWs.

3.3 Propagation in NWSIW

In this section the propagation of microwave signals is investigated through the description of the effective medium present in the waveguide, as influencing the propagation, and the formulation of the propagation constant and characteristic impedance of the waveguide.

3.3.1 Effective medium filling the NWSIW

The porous medium resulting from alumina anodization was illustrated in **Figure 2**. The (complex) permittivity of the equivalent effective medium, noted $\varepsilon_{\mathrm{AAO}}$, can be calculated using a simple volumetric law [20] involving porosity P of AAO and properties of bulk alumina reported in **Table 1**:

$$\varepsilon_{AAO} = P + (1 - P)\varepsilon_r\,(1 - i\,tan\delta) \tag{3}$$

As illustrated in **Figure 4**, both dielectric constant (left figure) and loss tangent factor (right figure) of porous AAO show a linear dependence on porosity P.

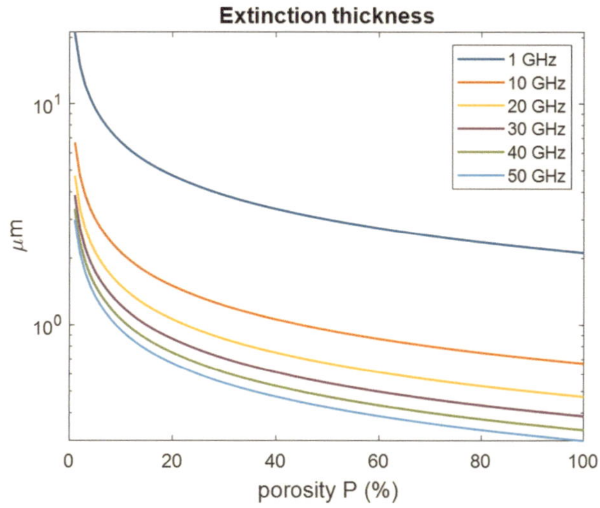

Figure 3.
Extinction thickness (Eq. (2)) as a function of porosity P.

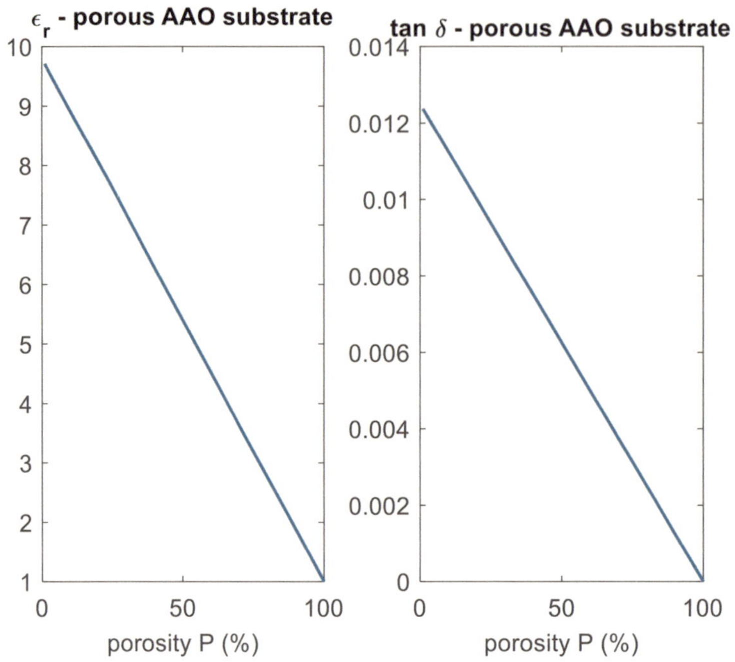

Figure 4.
Electrical properties of porous AAO depending on porosity P, according to Eq. (3). Left: Dielectric constant, right: Loss tangent factor.

For P = 0, the dielectric constant is that of bulk alumina given in **Table 1**, while for P = 100% it corresponds to air since no more alumina is present.

The same is true for the loss tangent factor; it goes from 0.0125 for P = 0%, corresponding to the value in **Table 1** for bulk alumina, to zero since for P = 100% the substrate reduces to air having no significant losses.

It has to be noted however that for practical use, AAO templates having high porosity should be avoided since they are much more brittle.

3.3.2 Propagation characteristics

The propagation inside a NWSIW is very similar to that occurring in a classical MRW. Given its width noted W and height equal to the thickness T of AAO template given in **Table 1**, the complex propagation constant, noted γ, is given by the following general expression [23]:

$$\gamma = \alpha + j\,\beta = \sqrt{\left(\frac{n\,\pi}{W}\right)^2 + \left(\frac{m\,\pi}{T}\right)^2 - \varepsilon_{AAO}\left(\frac{2\,\pi f}{c_o}\right)^2} \tag{4}$$

In Eq. (4) m and n are indices associated to TE_{mn} and TM_{mn} modes of propagation in an MRW. In the case of NWSIW, the thickness of the AAO being much lower than the width W of the guide, the first modes of propagation are TE_{m0} modes. Eq. (4) indeed reveals that the propagation constant presents a cut-off phenomenon; propagation occurs ($\beta > 0$) only above a certain frequency named cut-off frequency and noted f_c. Below f_c, the signal is attenuated instead of propagating ($\beta = 0$, $\alpha > 0$).

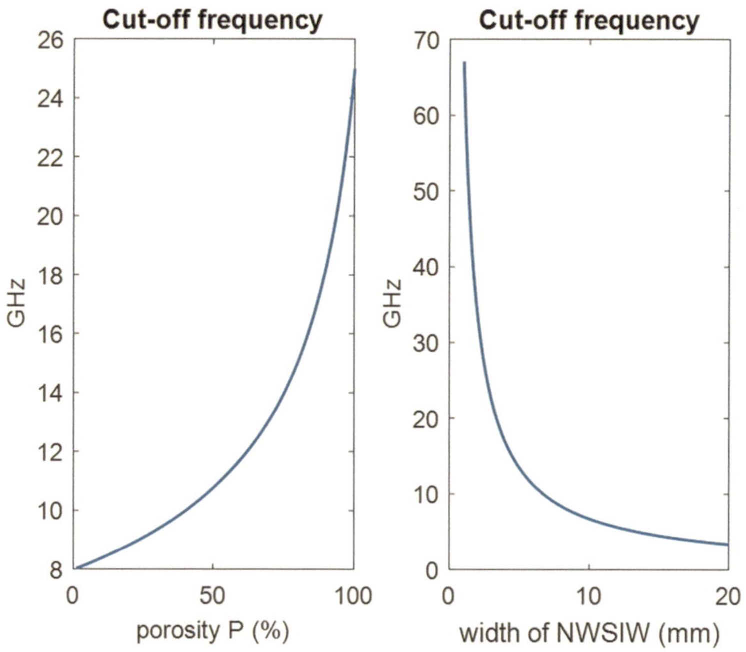

Figure 5.
Cut-off frequency f_{c10} according to Eq. (5). Left: Function of porosity P, for W = 6 mm. Right: Function of NWSIW width W, for P = 50%.

Derived from setting γ in Eq. (4) equal to zero, f_c writes as:

$$f_{c\,m0} = \frac{m\,c_o}{2\,W\,\varepsilon_{AAO}} \tag{5}$$

where W is the width of the waveguide and c_o the light velocity in air. For the NWSIW the three lowest cut-of frequencies occur for TE_{m0} modes.

Two statements can be derived. At first, the permittivity of AAO filling the waveguide depends on porosity P. This obviously influences the cut-off, hence the propagation constant, by virtue of (4–5). This illustrated in **Figures 5** and **6**, left. As porosity P increases, the dielectric constant decreases so that cut-off moves to higher frequencies (**Figure 5**), and this is reflected in the behavior of propagation constant (**Figure 6**). For a proper operation, that is allowing propagation in the 10–60 GHz range, porosity P should not exceed 40%.

Secondly, similar conclusions can be drawn as concerns the influence of the width of the waveguide. The cut-off frequency decreases as the width of the NWSIW increases, and a width superior to 5 mm is necessary for obtaining propagation in the NWSIW starting at 10 GHz.

3.3.3 Characteristic impedance of NWSIW

The characteristic impedance of a NWSIW has an expression similar to an MRW, expressed here as function of geometrical parameters W and T of the waveguide [23]:

$$Z_c = \frac{2\,T}{W}\,\frac{j\,2\pi f\,\mu_o}{\gamma} \tag{6}$$

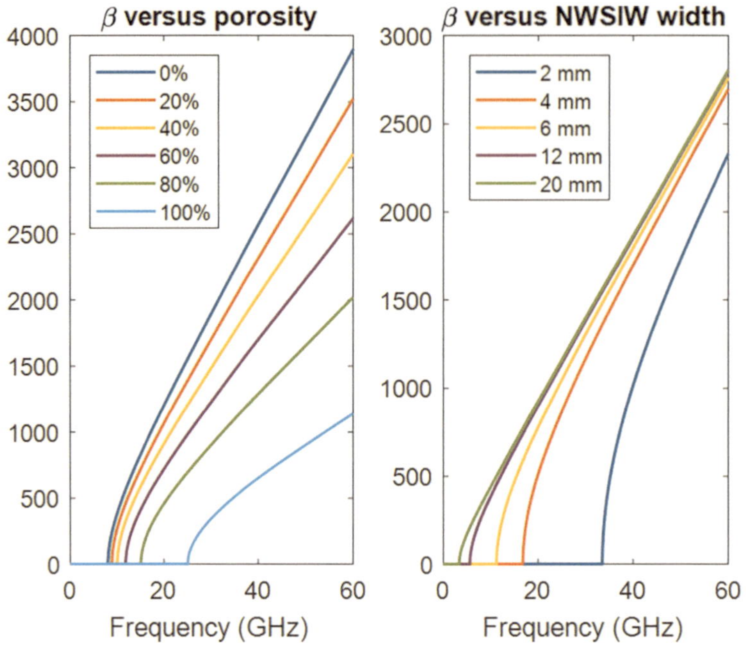

Figure 6.
Frequency dependence of propagation coefficient β given by Eq. (4). Left: For different values of porosity P, and W = 6 mm. Right: For different values of NWSIW width W, and P = 50%.

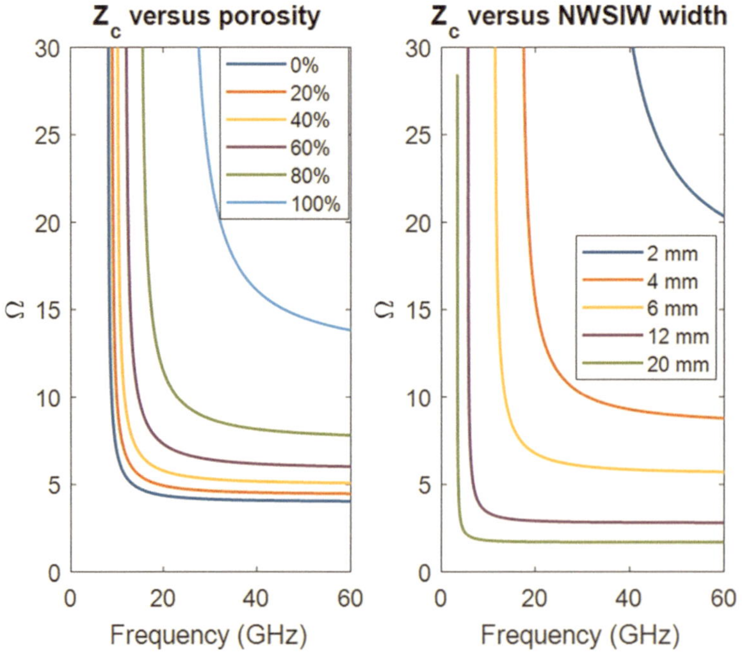

Figure 7.
Frequency dependence of characteristic impedance Z_c given by Eq. (6). Left: For different values of porosity P, and W = 6 mm. Right: For different values of NWSIW width W, and P = 50%.

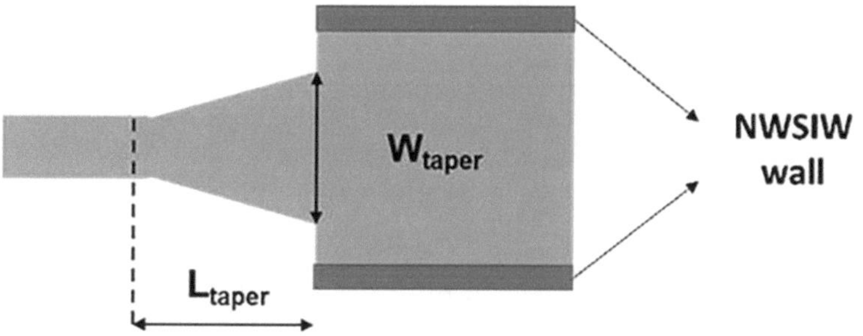

Figure 8.
Schematic representation of tapered microstrip-to-NWSIW transition for impedance matching.

Introducing expression (4) of γ into (6), the characteristic impedance Z_c can be represented in **Figure 7** as a function of frequency and for different values of porosity (left) and width of NWSIW (right). Resulting from definition (6) Z_c has a singularity at cut-off frequency, since at this frequency γ present in the denominator of (6) tends to zero.

Another important feature is the low level of characteristic impedance far above cut-off. Even for the most favorable conditions, i.e. high porosity in order to decrease the effective permittivity inside the waveguide (P = 80%), and moderate width of NWSIW (W = 4 mm), Z_c does not exceed 10 Ω in the flat constant regime above 20 GHz. As a result, a mismatch occurs with respect to the conventional 50 Ω reference impedance used for measurements of microwave devices.

3.3.4 Tapered transmission for improved matching

As a way to solve the problem of mismatch between the low impedance of the NWSIW and 50 Ω reference impedance, tapered transitions can be inserted between accesses of the NWSIW and microstrip or coplanar waveguides lines used for the connection with other devices. A schematic view of the microstrip-to-NWSIW transition is shown in **Figure 8**. The taper allows a smooth progressive change of characteristic impedance along taper length, which favors the matching.

Figure 9 shows simulated performances of microstripline -to - NWSIW transition for various values of taper length L = L_{taper} (left) and taper width W_{taper} (right).

Adequate values for maximizing transmission S_{21} and minimizing reflection S_{11} are L_{taper} = λ/4 where λ is the wavelength at the operating frequency assumed here equal to 10 GHz, and W_{taper} = 0.6 x W = 3.6 mm.

4. Fabrication techniques

Figure 10 shows the fabrication process for the NWSIW. As explained in Section 2, this technology differs from classical SIW since it takes advantage of the porous AAO template for creating by electrodeposition metallic nanowires inside the pores in order to form equivalent metallic shielding walls as explained in sections 3.1 and 3.2.

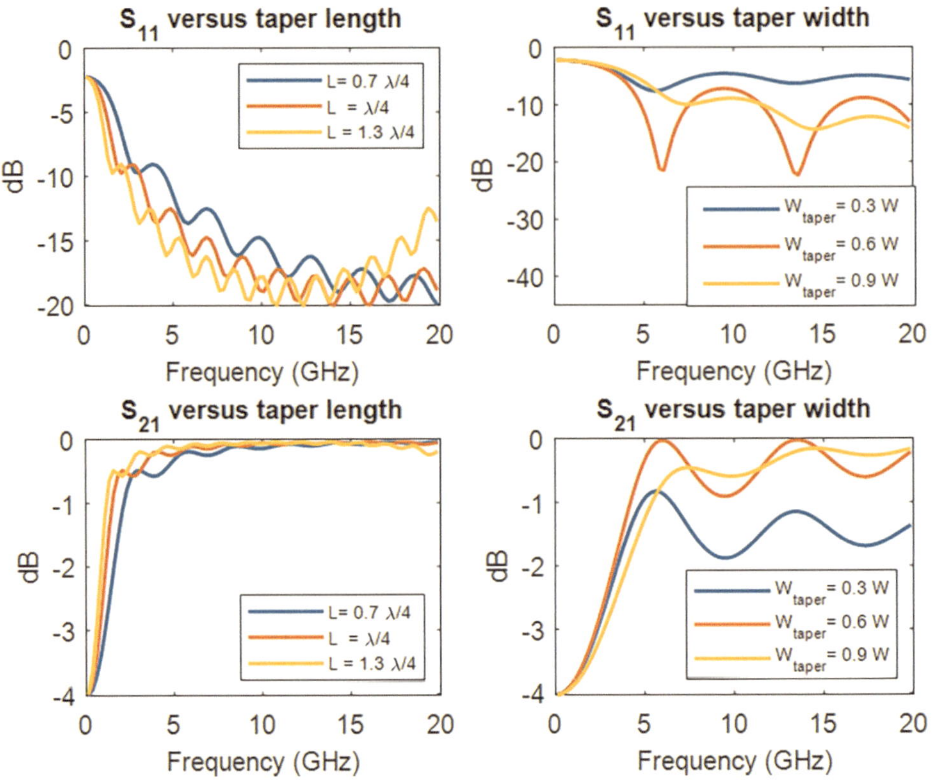

Figure 9.
*Performances of tapered transition of **Figure 8** as a function of taper length (left) and taper width (right). Top row: S_{11}, bottom row: S_{21}. Simulations are made for $W = 6$ mm and $P = 50\%$.*

The first step is the deposition of a metallic layer (blue color) by electroplating on the back side of the AAO template, step 1 in **Figure 10**; this layer is used as cathode during the electrodeposition process.

Then in step 2 the areas where electrodeposition of NW is needed are defined; equivalently the areas where electrodeposition is unwanted are determined. As detailed in [19–21] two options are possible, named A and B. In option A, a metal layer is electroplated on top of AAO. Openings are created by laser etching this top metal layer in areas where electrodeposition of MNW is desired, to allow the penetration of the electrolytic solution inside pores that are not covered by the metal. Option B is dual/opposite; the areas where MNW are unwanted are selected by laser burning locally the surface of the AAO in order to destroy locally the porosity and clog the pores, making electrodeposition of MNWimpossible in these areas. The burned surface of AAO is represented as a magenta layer in **Figure 3B**. It remains present during step 3 to 5. The result of step 2 for both option A and B is step 3, where only open pores are available for electrodeposition. The result of step 2 for both option A and B is step 3, where only open pores are available for electrodeposition.

The last 2 steps are similar for option A and B. In step 4, electrodeposition process described in [21] is used to grow MNW in free pores in such a way that pores are slightly overfilled. By doing so, during step 5 when electroplating of top

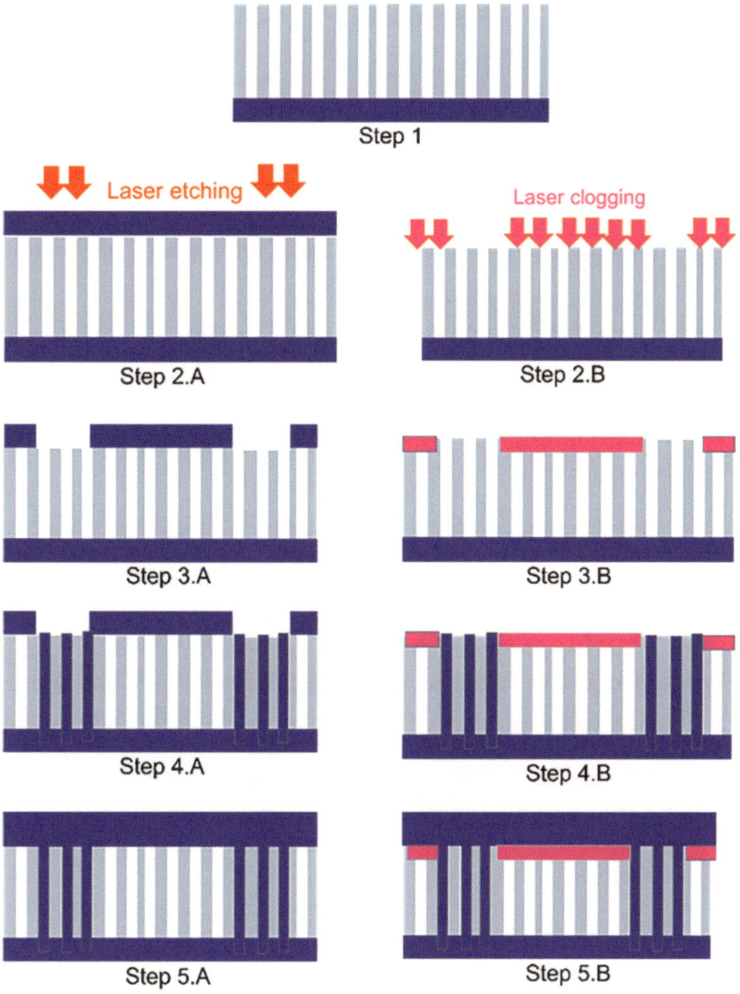

Figure 10.
Steps of fabrication process for the NWSIW.

metal layer occurs, good electrical contacts are created between top metal layer and upper end of MNW, forming efficient metallic shielding walls for the equivalent waveguide.

5. Practical realizations

5.1 NWSIW transmission line

Figure 11 shows the transmission measured in a NWSIW transmission line having a width W = 6 mm, and built on a 100 μm-thick AAO template. The location of the three first cut-off frequencies corresponding to TE_{10}, TE_{20} and TE_{30} modes are visible; their values are in good correspondence with Eq. (5). For a proper operation of NWSIW devices, they should be designed for the 8–18 GHz frequency range, i.e. between the first and second cut-off frequencies.

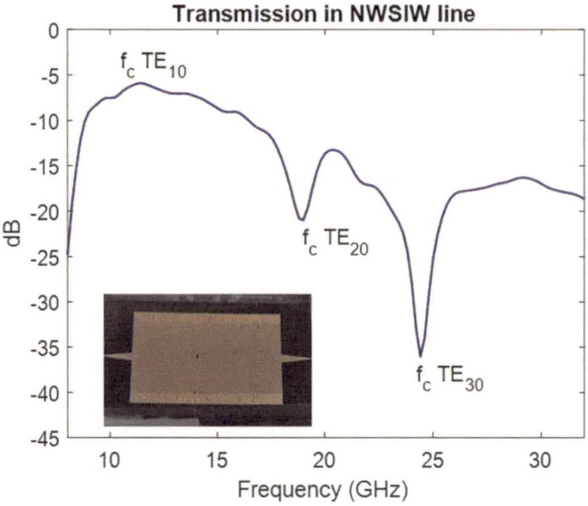

Figure 11.
Transmission measured on a NWSIW built in a 100 µm-thick AAO template. Inset shows picture of the device having width W = 6 mm.

Insert of **Figure 11** shows a picture of the fabricated NWSIW line. Material for MNW is Cu and areas filled by MNW are visible as lighter strips on upper and lower side of the brown rectangle forming the top metallization of the NWSIW, and having a length equal to 12 mm. On left- and right-hand sides of the picture taper microstrip transitions aiming at improving impedance matching are visible.

5.2 NWSIW isolator

Figure 12 shows a practical realization of a NWSIW isolator basing on the operation mode of a ferrite-slab resonant mode isolator in classical MRW technology [24, 25]. Here in order to mimick the ferrite slab, a narrow wall made of ferromagnetic NW (alloy of Nickel and iron, noted NiFe, red color) is grown close to a wall of the NWSIW (blue color), made of Cu NW grown in AAO, as shown in the inset of **Figure 12**. It has to be noted that NiFe NW forming the slab are not grown over the full height of AAO template in order to avoid the formation of a short-circuit due to contacts between conductive NiFe NW and top metallization of the waveguide, since a short-circuit would prevent the transmission of the signal.

As expected a nonreciprocal transmission is observed in the NWSIW, due to the presence of ferromagnetic material located at a specific position in the waveguide. Transmission S_{21} in forward direction is more than 10 dB lower than transmission S_{12} in reverse direction. We can conclude that the device ensures an isolation level superior to 10 dB [21], which is close to the state-of-the art, as shown in **Table 2**.

Compared to isolators based on ferrite slabs in MRW (resonant-mode isolator), the planar devices are much thinner and usually do not require an external magnetic field to bias the ferrite. The best result is obtained by Cheng et al. [26]: their planar ferrite resonance isolator is 635 µm-thick and works without DC magnetic field bias. Hemour et al. [27] propose an SIW isolator based on ferrite, with lower performances reported in **Table 2**.

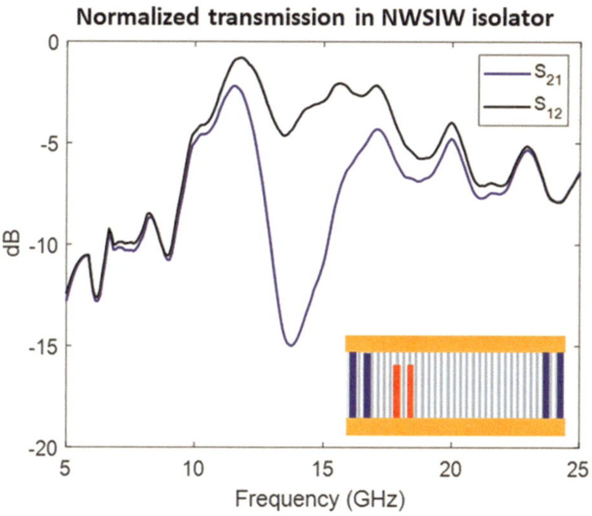

Figure 12.
Measurement of NWSIW isolator. Inset shows the topology: Cu NW (blue) form lateral walls, while NiFe NW (red) form a ferromagnetic slab. Electroplated copper metal layers (orange) form bottom and top walls of NWSIW.

Reference	Topology/ technology	Material	F (GHz)	Isolation (dB/cm)	Insertion losses (dB)	Thickness (mm)
[26]	SIW in PCB	Ferrite slab	10	20	3	0.635
[27]	SIW in PCB	Ferrite slab	11	8	7	na
[28]	Microstrip on PC	Co NW	41	7	15	0.021
[29]	CPW on AAO	Ni NW	20	6.5	na	0.06
[13]	Microstrip on AAO	CoFe NW	24	14	10	0.20
[18]	NWSIW on AAO	NiFe NW	13	7	na	0.05
Figure 12 [21]	NWSIW on AAO	NiFe NW	12	11	na	0.10

Table 2.
Comparison of performances of planar SIW isolators.

The first demonstration of an isolator based on ferromagnetic nanowires (NW) was reported in [28]. Cobalt NW are included in nanoporous polycarbonate (PC) thin substrate, and 10 dB isolation is obtained, however with 15 dB insertion losses and requiring a 9 kOe DC magnetic bias. Next two other devices based on NW in AAO porous template are reported in the literature. Kuarn [29] proposed a coplanar waveguide (CPW) technology including Nickel (Ni) MNW in AAO. Reported isolation level is 6 dB/cm at 23 GHz under 5.6 kOe DC magnetic bias. The best performances for nanowire-based planar isolators were obtained by Carignan et al. [13], who measured an isolation of 14 dB/cm without DC bias on a microstrip topology including CoFeB in AAO.

Two realizations of UCLouvain combining NW in AAO and planar SIW into an NWSIW isolator are reported in the literature and do not require external DC bias.

The first one [18] shows an isolation level of 7 dB/cm, while the second one [21] shown in **Figure 12** has an isolation of 11 dB/cm. These two last realizations are among the thinnest according to **Table 2**.

5.3 NWSIW EBG filter

As last illustration in this chapter, **Figure 13** presents the topology of the so-called EBG filter (for Electromagnetic Band Gap) realized in NWSIW technology. The EBG effect associated to periodic structures was introduced in the '90s [11, 30–32]. The filter is based on the cascade of three NWSIW sections partially filled with Nickel MNW and separated by empty NWSIW sections.

Measurement of transmission in **Figure 14** shows that a gap in the transmission occurs at 20 GHz. It is created by the contrast between the dielectric constant of AAO

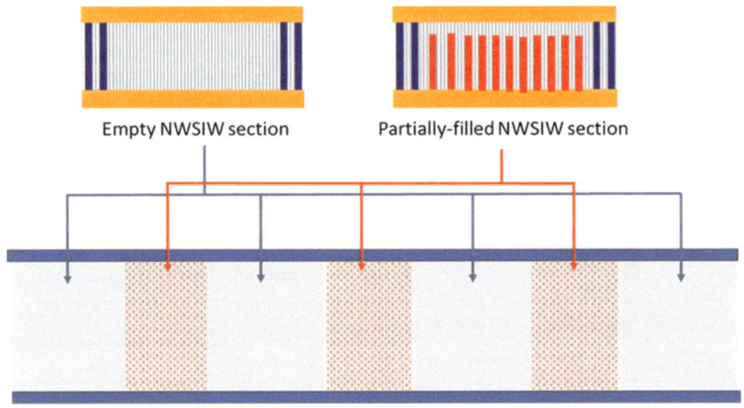

Figure 13.
Topology of EBG filter in NWSIW technology. Blue NW form vertical walls of NWSIW, while red NW having filling height h = 75% are grown in the NWSIW in order to create the EBG effect. Electroplated copper metal layers (orange) form bottom and top walls of NWSIW.

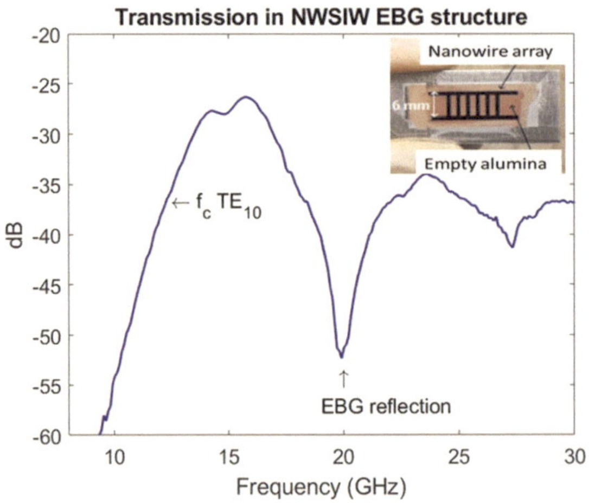

Figure 14.
Measured transmission in EBG filter in NWSIW technology. Inset: Photograph before electroplating realized at step 5 of ***Figure 10****.*

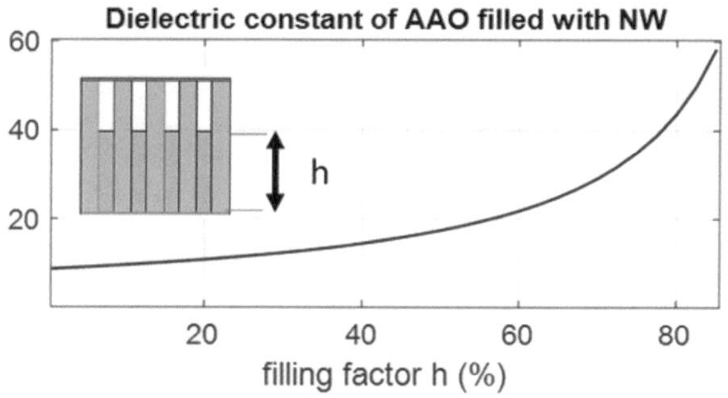

Figure 15.
Influence of filling height of NW on dielectric constant.

empty sections and the much higher value of dielectric constant for sections partially filled with Ni NW over a relative filling factor h. The behavior of the dielectric constant versus h is shown in **Figure 15** and was initially introduced in [15]. The generation of the bandgap at a dedicated frequency f_o requires that the length of each section, empty and filled, is equal to a quarter wavelength at f_o. This condition is expressed by equations (7–8):

$$L_{empty\ section} = \frac{c_o}{4\,f_o\,\varepsilon_{r\ empty}} = 2.6\ \text{mm}. \tag{7}$$

$$L_{filled\ section} = \frac{c_o}{4\,f_o\,\varepsilon_{r\ filled}} = 5.1\ \text{mm} \tag{8}$$

The realized filter has a filling factor h = 75% yielding 34 as value for $\varepsilon_{r\ filled}$, while dielectric constant of empty AAO with P = 4% is close to 9.8. The contrast between dielectric constant of filled and empty sections is high and responsible for the bandgap at f_o = 20 GHz. The depth of the bandgap is −50 dB, which, compared to the maximal transmission level of - 25 dB, means a stopband rejection effect of 25 dB.

6. Challenges and perspectives of the new NWSIW technology

Coming to the end of this chapter, it is now time to conclude on the performances, the maturity and the perspectives of the NWSIW technology.

The NWSIW presented in this chapter and the classical planar SIW of **Figure 1** share about the same pros and cons:

- A fully planar geometry, based on a thin planar substrate, much more compact than classical MRW
    - The resulting disadvantage is that both topologies require specific matching structures to solve the impedance mismatch issue resulting from the low thickness of the substrate.
    - However, the issue disappears if a whole architecture of devices is built in the same NWSIW technology on a single AAO substrate. The design can be made for a same reference impedance, so that a single transition will be needed for

Reference	Topology	Technology	f (GHz)	Insertion losses (dB/cm)
[33]	CPW	NixSiy/Si	10	2
[34]	CPW	MMIC	20	5
[35]	SIW	LTCC	50	2.5
[36]	SIW	LTCC	38	4
[37]	Microstrip	NW in AAO	60	7
[38]	SIW	NW in AAO	20	7.4
[20]	NWSIW	NW in AAO	12	4.4

Table 3.
Comparison of insertion loss.

interfacing the designed circuit in NWSIW technology with 50 Ω reference impedance.

- An improved immunity to radiation and crosstalk interferences compared to open planar structures
    - For classical SIW, vias have to be carefully dimensioned and spaced in order to induce an efficient shield.
    - while for NWSIW a thin wall of grown nanowires is sufficient to form an efficient shield, since the nanoporous AAO template imposes a close packing of the electrodeposited MNW array.
- Compared to MRW that are empty, both SIW and NWSIW structure are filled by a medium having permittivity higher than air.
    - This reduces the size of the device for a given operational frequency
    - Devices have to cope with dielectric loss of the substrate filling the waveguide.

However, the insertion losses observed in **Figure 11** for NWSIW (8 dB at 12 GHz, corresponding to 4.4 dB/cm) are comparable or lower than other technologies/topologies, as shown **Table 3**. Insertion losses should be significantly decreased by further improvement of fabrication process. Indeed, the upper and lower Cu layers electroplated on the two faces of the AAO membrane can be thickened in order to attenuate the effect of the roughness of the metallization due its imperfect contact with porous template during step 1 and 5 of the fabrication (**Figure 10**). This will increase the shielding effect while reducing ohmic losses, in order to achieve insertion losses lower than 1 dB/cm.

The road to achieving these objectives is not actually that long. We have shown that the combination of nanowires and SIW technologies is meaningful, since it allows a facile integration of various functionalities such as filtering and nonreciprocity on a same substrate thanks to the growth of various kinds of MNW in AAO while tuning their filling height. The fabrication of a platform of miniaturized NWSIW devices that rival the state-of-the-art will be made possible by some optimizations of the fabrication process as outlined above.

Acknowledgements

The authors would like to thank Ester Tooten, Pascal Simon and Pascal Van Velthem for their help and advices.

This work was supported by the Research Science Foundation of Belgium (FRS-FNRS). Vivien VanKerckhoven acknowledges the FRS-FNRS for financial support (FRIA grant). Isabelle Huynen is Research Director of FRS-FNRS.

Nomenclature

The following abbreviations are used in this manuscript:

AAO	Anodic Aluminum Oxide
CCL	copper clad laminate
CPW	Coplanar Waveguide
Cu	Copper
EBG	Electromagnetic Band Gap
NiFe	nickel-iron alloy
MMIC	Monolithic Microwave Integrated Circuit
MNW	metallic nanowire
LTTC	Low Temperature Co-fired Ceramics
NW	Nanowire
NWSIW	Nanowired Substrate Integrated Waveguide
PCB	Printed Circuits Boards
SIW	Substrate Integrated Waveguide
UCLouvain	Université catholique de Louvain

Author details

Vivien Van Kerckhoven, Luc Piraux and Isabelle Huynen*
Université catholique de Louvain (UCLouvain), Louvain-la-Neuve, Belgium

*Address all correspondence to: isabelle.huynen@uclouvain.be

IntechOpen

© 2022 The Author(s). Licensee IntechOpen. This chapter is distributed under the terms of the Creative Commons Attribution License (http://creativecommons.org/licenses/by/3.0), which permits unrestricted use, distribution, and reproduction in any medium, provided the original work is properly cited.

References

[1] Wu K, Deslandes D, Cassivi Y. The substrate integrated circuits - A new concept for high-frequency electronics and optoelectronics. In: Proceedings of the 6th International Conference on Telecommunications in Modern Satellite, Cable and Broadcasting Service 2003—TELSIKS 2003, Nis, Yugoslavia, 1–3 October 2003. USA: IEEE; 2003. pp. III-X. DOI: 10.1109/TELSKS.2003.1246173

[2] Pazian M, Bozzi M, Perregrini L. Crosstalk in substrate integrated waveguides. IEEE Transactions on Electromagnetic Compatibility. 2015;**57**: 80-86. DOI: 10.1109/TEMC.2014.2364634

[3] Simpson JJ, Taflove A, Mix JA, Heck H. Substrate integrated waveguides optimized for ultrahighspeed digital interconnects. IEEE Transactions on Microwave Theory and Techniques. 2006;**54**:1983-1990. DOI: 10.1109/TMTT.2006.873622

[4] Lai QH, Fumeaux C, Hong W. Mutual coupling between parallel half-mode substrate integrated waveguides. In: Proceedings of the 2012 International Conference on Microwave and Millimeter Wave Technology (ICMMT), Shenzhen, China, 5–8 May 2012. USA: IEEE; 2012. p. 4. DOI: 10.1109/ICMMT.2012.6230369

[5] Bozzi M, Georgiadis A, Wu K. Review of substrate-integrated waveguide circuits and antennas. IET Microwaves, Antennas and Propagation. 2011;**5**: 909-920. DOI: 10.1049/iet-map.2010.0463

[6] Moro R, Bozzi M, Collado A, Georgiadis A, Via S. Plastic-based substrate integrated waveguide (SIW) components and antennas. In: Proc. Eur. Microwave Integr. Circuits Conf. Amsterdam, The Netherlands, 29 Oct.-1 Nov. 2012. USA: IEEE; 2012. pp. 627-630. DOI: 10.23919/EuMC.2012.6459177

[7] Kuanr B, Veerakumar V, Marson R, Mishra SR, Kuanr AV, Camley RE, et al. Nickel nano-wires filled alumina templates for microwave electronics. IEEE Transactions on Magnetics Letters. 2009;**45**:4052-4055. DOI: 10.1109/TMAG.2009.2024880

[8] Qiao L, Han X, Gao B, Wang J, Wen F, Li F. Microwave absorption properties of the hierarchically branched Ni nanowire composites. Journal of Applied Physics. 2009;**105**:053911. DOI: 10.1063/1.3081649

[9] Chen W, Han M, Deng L. High frequency microwave absorbing properties of cobalt nanowires with transverse magnetocrystalline anisotropy. Physica B: Condensed Matter. 2010;**405**:1484-1488. DOI: 10.1016/j.physb.2009.12.026

[10] Hamoir G, Piraux L, Huynen I. Q-factor improvement of integrated inductors using high aspect ratio ferromagnetic nanowires. Microwave and Optical Technology Letters. 2012;**54**: 1484-1488. DOI: 10.1002/mop.26900

[11] Saib A, Vanhoenacker-Janvier D, Huynen I, Encinas A, Piraux L, Ferain E, et al. Magnetic photonic band-gap material at microwave frequencies based on ferromagnetic nanowires. Applied Physics Letters. 2003;**83**:2378-2380. DOI: 10.1063/1.1610798

[12] Hamoir G, De La Torre MJ, Piraux L, Huynen I. Self-biased nonreciprocal microstrip phase shifter on magnetic

nanowired substrate suitable for gyrator applications. IEEE Transactions on Microwave Theory and Techniques. 2012;**60**:152-157. DOI: 10.1109/TMTT.2012.2195016

[13] Carignan L, Caloz C, Menard D. Dual-band integrated self-biased edge-mode isolator based on the double ferromagnetic resonance of a bistable nanowire substrate. In: Microwave Theory and Techniques, Proceedings of IEEE Symposium Digest. Anaheim, CA, USA: IEEE; 2010. pp. 1336-1339. DOI: 10.1109/MWSYM.2010.5517585

[14] Saib A, Darques M, Piraux L, Vanhoenacker-Janvier D, Huynen I. An unbiased integrated microstrip circulator based on magnetic nanowired substrate. IEEE Transactions on Microwave Theory and Techniques. 2003;**83**:237. DOI: 10.1109/TMTT.2005.848818

[15] Spiegel J, De La Torre J, Darques M, Piraux L, Huynen I. Permittivity model for ferromagnetic nanowired substrates. IEEE Microwave and Wireless Components Letters. 2007;**60**: 492-494. DOI: 10.1109/LMWC.2007.899303

[16] Ramirez-Villegas R, Huynen I, Piraux L, Encinas A, De La, Torre MJ. Self-biased nonreciprocal microstrip phase shifter on magnetic nanowired substrate suitable for gyrator applications. IEEE Transactions on Microwave Theory and Techniques. 2016;**65**:72-77. DOI: 10.1109/TMTT.2012.2195016

[17] Van Kerckhoven V, Piraux L, Huynen I. A novel laser-assisted fabrication process for nanowired substrate integrated devices. In: 2018 13th European Microwave Integrated Circuits Conference (EuMIC). 23-25 Sept. 2018, Madrid, Spain. USA: IEEE; 2018. pp. 170-173. DOI: 10.23919/EuMIC.2018.8539911

[18] Van Kerckhoven V, Piraux L, Huynen I. Substrate integrated waveguide isolator based on ferromagnetic nanowires in porous alumina template. Applied Physics Letters. 2014;**105**:183107. DOI: 10.1063/1.4901277

[19] Van Kerckhoven V, Piraux L, Huynen I. A laser-assisted process to produce patterned growth of vertically aligned nanowire arrays for monolithic microwave integrated devices. Nanotechnology. 2016;**65**:72-77. DOI: 10.1088/0957-4484/27/23/235301

[20] Van Kerckhoven V, Piraux L, Huynen I. Design of substrate integrated waveguides based on nanowires: Numerical guidelines. International Journal of Numerical Modelling. 2021; **34**:e2906. DOI: 10.1002/jnm.2906

[21] Van Kerckhoven V, Piraux L, Huynen I. Fabrication of microwave devices based on magnetic nanowires using a laser-assisted process. Micromachines. 2019;**10**:475. DOI: 10.3390/mi10070475

[22] Van Kerckhoven V. Nanowire-Based Microwave Devices in Substrate Integrated Waveguide Topology Using a Laser-Assisted Fabrication Process. Louvain-La-Neuve: UCLouvain; 2019

[23] Pozar DM. Microwave Engineering. 3rd ed. Hoboken. USA: John Wiley & Sons; 2005. p. 756. DOI: 10.1002/9780470974704

[24] Schloemann E. On the theory of the ferrite resonance isolator. IRE Transactions on Microwave Theory and Techniques. 1960;**199**:605-611. DOI: 10.1109/TMTT.1960.1124723

[25] Comstock RL, Fay CE. Operation of the field displacement in rectangular waveguide isolator. IRE Transactions on Microwave Theory and Techniques. 1960;**199**:199-206. DOI: 10.1109/TMTT.1960.1124804

[26] Cheng YJ, Huang QD, Wang YR, Li JW. Narrowband substrate integrated waveguide isolators. IEEE Microwave and Wireless Components Letters. 2014; **24**:698-700. DOI: 10.1109/LMWC.2014.2344440

[27] Hemour S, Adhikari S, Ghiotto A, Wu K. Low magnetic biased SIW-based isolator: Effect of the rising temperature on the performance of the isolator. In: Proceedings of the WAMICON Conference. 6 June 2014. Tampa, USA: IEEE; 2014. p. 3. DOI: 10.1109/WAMICON.2014.6857801

[28] Spiegel J, De la Torre J, Piraux L, Huynen I. Isolator concept based on ferromagnetic nanowired substrates. In: Proceedings of International Microwave Symposium (IMS2009), June 7–12, 2009. Boston, USA: IEEE; 2006. pp. 29-32. DOI: 10.1109/MWSYM.2009.5165624

[29] Kuanr BK, Veerakumar V, Marson RL, Mishra SR, Carnley RE, Celinski Z. Nonreciprocal microwave devices based on magnetic nanowires. Applied Physics Letters. 2009;**94**: 202505. DOI: 10.1063/1.3124657

[30] Yablonovitch E. Inhibited spontaneous emission in solid-state physics and electronics. Physical Review Letters. 1987;**58**:2059. DOI: 10.1103/PhysRevLett.58.2059

[31] Laso MAG, Erro MJ, Benito D, Garde MJ, Lopetegi T, Falcone F, et al. Analysis and design of 1-D photonic bandgap microstrip structures using a fiber grating model. Microwave and Optical Technology Letters. 1999;**22**: 223-226. DOI: 10.1002/(SICI)1098-2760(19990820)22:4< 223: AID-MOP1>3.0.CO;2-X

[32] Sigalas MM, Soukoulis CM, Biswas R, Ho KH. Effect of the magnetic permeability on photonic band gaps. Physical Review B. 1997;**56**:959. DOI: 10.1103/PhysRevB.56.959

[33] Raman S, Rucky F, Rebeiz G. A high-performance W-band uniplanar subharmonic mixer. IEEE Transactions on Microwave Theory and Techniques. 1997; **45**:955-959. DOI: 10.1109/22.588609

[34] Kim J, Qian Y, Feng G, Ma P, Judy J, Chang MF, et al. A novel low-loss low-crosstalk interconnect for broad-band mixed-signal silicon MMIC's. IEEE Transactions on Microwave Theory and Techniques. 1999;**47**:1830-1834. DOI: 10.1109/22.788519

[35] Dancila D, Rottenberg X, Tilmans HAC, De Raedt WR, Huynen I. 57-64 GHz seven-pole bandpass filter substrate integrated waveguide (SIW) in LTCC. In: Proceedings of the 2011 IEEE MTT-S International Microwave Workshop on Millimeter Wave Integration Technologies, 15–16 September 2011; Sitges, Spain. USA: IEEE; 2011. pp. 200-203. DOI: 10.1109/IMWS3.2011.6061876

[36] Hizan HM, Ambak Z, Ibrahim A, Yusoff MZM, Kanesan T. Effect of insertion losses on millimeter-wave SIW filters using LTCC technology. In: Proceedings of the 2015 IEEE International RF and Microwave Conference (RFM 2015); 14–16 December 2015, Kuching, Malaysia. USA: IEEE; 2015. pp. 64-66. DOI: 10.1109/RFM.2015.7587714

[37] Bertrand M, Rehder GP, Serrano ALC, Gomes LG, Pinheiro JM, Alvarenga RCA, et al. Integrated

waveguides in nanoporous alumina membrane for millimeter-wave interposer. IEEE Microwave and Wireless Components Letters. 2019;**29**: 83-87. DOI: 10.1109/ LMWC.2018.2887193

[38] Hamoir G. Microwave Devices Based on Unbiased Tunable Ferromagnetic Nanowire Arrays. Louvain-la-Neuve: UCLouvain; 2014

Chapter 3

Novel Filtering Applications in Substrate-Integrated Waveguide Technology

Angela Coves and Maurizio Bozzi

Abstract

The SIW technology combines complete shielding and fairly low losses with simple and cost-effective manufacturing, thus representing the ideal platform for the development of the next generation of wireless systems, including the band-pass filters among them. In this chapter, a number of novel SIW filter configurations will be presented to improve the filter performance, reduce losses, and minimize the filter footprint. To this end, different topologies of band-pass filters in SIW technology will be described based on stepped-impedance configurations (with high and low dielectric constant sections) making use of the impedance inverter model, extending this concept to half-mode SIW structures, with the aim to reduce the size of the filters.

Keywords: filters, substrate-integrated waveguide (SIW), effective permittivity

1. Introduction

Substrate-integrated waveguides (SIWs) are planar structures that emulate a dielectric-filled rectangular waveguide (RWG) in a single circuit board, in which the lateral metallic walls are replaced with a periodic array of metallic vias (see **Figure 1**) [1, 2]. Thus, SIWs are good candidates to be used as building blocks for the implementation of microwave waveguide filters with different topologies, benefiting from the advantages of such technology (mainly low cost and easy integration), combined with the well-known advantages of conventional rectangular waveguides (complete shielding and high-power-handling capability).

In the following sections, we begin analyzing the main properties of ordinary SIWs with the homogeneous substrate, and those whose substrate is periodically loaded with either cylindrical air holes or with metallic cylinders, thus achieving a reduced/higher effective permittivity, respectively. After that, different topologies of band-pass filters in SIW technology are briefly described, starting from classical iris-type SIW filters and moving to more novel topologies, consisting of step impedance filters based on high and low dielectric constant sections, extending this concept to half-mode SIW structures, with the aim to reduce the size of the filters, showing in all cases good performances in terms of insertion and return losses in their passbands, along with deep and wide rejection bands.

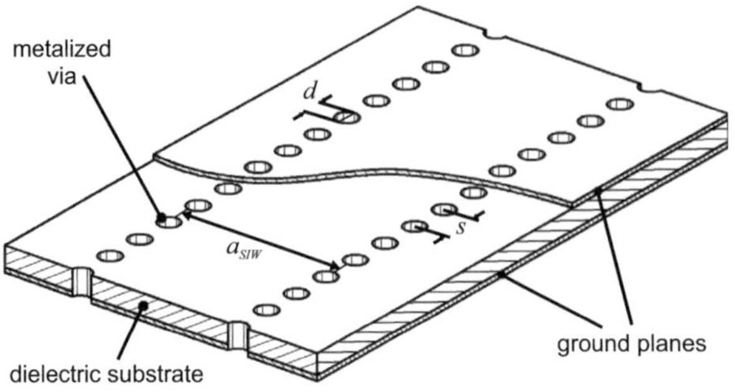

Figure 1.
The geometry of the classical SIW structure.

2. Ordinary SIW and SIW periodically loaded with cylindrical air holes or with metallic cylinders

SIWs are planar structures that emulate dielectric-filled rectangular waveguides (see **Figure 1**). The two ground planes represent the top and bottom metal walls of the rectangular waveguide, and the rows of metal vias replace the sidewalls of the waveguide. The ordinary SIW made with a homogeneous substrate basically has the same guided-wave characteristics as the conventional rectangular waveguide [3], and its fundamental mode is similar to the TE_{10} mode of the rectangular waveguide. However, since the electric current density on the metal vias can only flow in the vertical direction, only TE_{n0} modes are supported by SIWs [4]. The metallic vias are characterized by their separation s and diameter d. Their values must be appropriately chosen [5] to avoid radiation losses, so they must fulfill the following conditions:

$$d < \lambda_g/5, \quad s \leq 2d \tag{1}$$

where λ_g is the guided wavelength. The propagation constant of the SIW fundamental mode is mainly determined by the width a_{SIW} of the SIW (see **Figure 1**). A previous study [3] demonstrates that a SIW can be analyzed as an equivalent rectangular waveguide of effective width a given by

$$a = a_{SIW} - \frac{d^2}{0.95\,s}. \tag{2}$$

Therefore, all the presented results in this section have been obtained using the equivalent waveguide of width a, related to the cutoff frequency f_{c10} of the TE_{10} mode and the relative permittivity of the substrate material by:

$$a = \frac{c}{2f_{c10}\sqrt{\varepsilon_r}}. \tag{3}$$

Going a step further, some propagation regions of the SIW can be conveniently modified so that it can behave as if it is loaded by a different dielectric permittivity in such regions with respect to that of its substrate, which may be of practical interest in

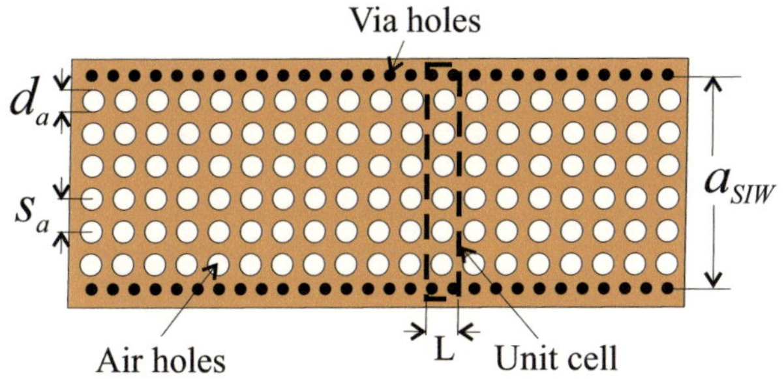

Figure 2.
Scheme of a SIW with periodic air holes with a rectangular pattern.

filtering applications, as shown in the following sections. A simple way of it can be achieved when the propagation region of the SIW is periodically loaded by air holes (see **Figure 2**), in which case a considerable reduction of the effective permittivity can be obtained (as long as the perforated substrate is shielded in the top and bottom walls, so the electric field distributes transversely through both the substrate and the air holes regions following the TE_{10} mode profile). Additionally, it is expected a decrease in dielectric losses due to the removal of substrate material, which may be of special interest in high-frequency bands. Effective permittivity of the periodically perforated SIW can be obtained to be used in filtering design by analyzing a unit cell of the perforated structure, by using the eigenmode solver of the commercial software tool Ansys HFSS [6], so that it is possible to relate the effective permittivity of the waveguide with the cutoff frequency of the TE_{10} mode through the following expression:

$$\varepsilon_r = \frac{c^2}{4a^2 f_{c10}^2} \tag{4}$$

A parametric study of the effective permittivity obtained in a SIW periodically loaded with cylindrical air holes following a rectangular pattern can be found in Ref. [7], where it has been analyzed the effect of the air holes parameters (the diameter d_a and separation s_a) in the resultant effective relative permittivity of the waveguide, achieving a reduction of more than a 60% of the substrate relative permittivity.

Alternatively, the use of high effective permittivity structures, which behave as slow-wave structures, are also of special interest for device miniaturization in filtering applications. With this regard, the increase of the effective permittivity of a SIW can also be obtained by inserting in the dielectric an array of metallic inclusions, as already demonstrated in Refs. [8, 9]. A simple implementation of a SIW with a high effective permittivity can be achieved by inserting an array of metallic cylinders (see **Figure 3**), whose height must be lower than but not far from the waveguide height, to achieve a high effective permittivity in the waveguide. In Ref. [10], a parametric study of the effective permittivity in a SIW has been done, in which an array of metallic cylinders with a triangular pattern has been inserted to synthesize a higher effective permittivity, obtaining an effective permittivity that is more than twice the value of the substrate permittivity with the proper selection of the cylinder parameters (the diameter d_c and separation s_c).

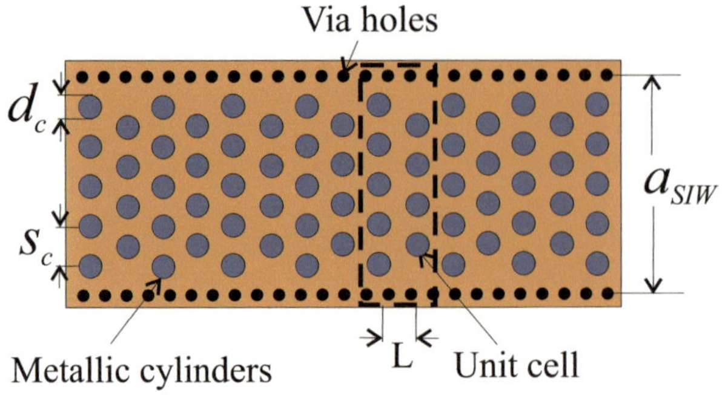

Figure 3.
Scheme of a SIW with an array of periodic cylinders with a triangular pattern.

Both types of periodically loaded SIWs with reduced or increased effective permittivity can be employed in novel filter solutions, as shown in the following section.

3. Band-pass step impedance filters in SIW technology

In this section, the design procedure of novel band-pass filters in SIW technology using the impedance inverter model has been described [11], where the quarter-wave sections constituting the resonators are coupled through evanescent mode sections, which can be implemented by classical iris waveguides, or alternatively, by reduced permittivity SIW sections.

3.1 Band-pass iris filter in SIW technology

We begin this section by giving the basic guidelines for the practical design of a waveguide iris filter in SIW technology using the well-known impedance inverter model [11], which will be used as the basis for more sophisticated stepped-impedance configurations of SIW filters (with high and low dielectric constant sections). The filter design is made using the equivalent rectangular waveguide of effective width a given in Eq. (2), and the final design of the equivalent filter in SIW technology is accomplished using such equivalence in each of the respective waveguide sections. For the design of the waveguide-based filter, the equivalent circuit model of impedance inverters of an inductive waveguide iris through a T network can be employed [11], as can be seen in **Figure 4(a)**. The filter consists of half-wave resonators separated by the inductive iris. Using an electromagnetic simulator, the iris scattering matrix can be obtained and therefore its equivalent T network. Each iris is represented by two series reactances denoted by X_s and a shunt reactance denoted by X_p. The equivalent circuital rectangular iris filter is shown in **Figure 4(a)**. To transform it into the impedance inverter model, we use the impedance inverter circuit consisting of an inductive T network and two sections of length $\varphi/2$ on each side. The inverter is created by adding a length $\varphi/2$ and $-\varphi/2$ on each side of the discontinuity, as shown in **Figure 4(b)**. In this case, the resonators are transmission lines of length L_n connected to two transmission lines of artificial lengths $-\varphi_n/2$ and $-\varphi_{n+1}/2$. These

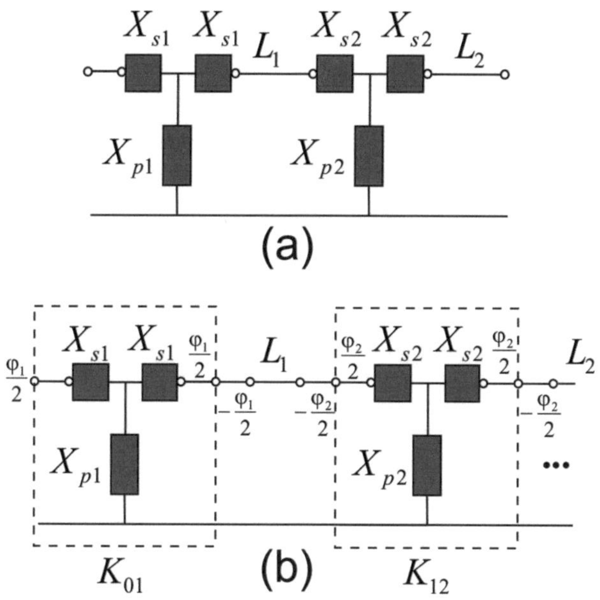

Figure 4.
(a) Equivalent circuit model of an inductive waveguide iris through a T *network. (b) Equivalent impedance inverters model of an inductive waveguide iris through a* T *network.*

lengths represent the load of the resonator from the adjacent coupling inverters (**Figure 4(b)**).

As an example, we show the design of a five-pole Chebyshev filter, as illustrated in **Figure 5**, consisting of several sections of rectangular waveguide coupled with an inductive iris. The filter is designed with a center frequency $f_0 = 4$ GHz, a bandwidth of 600 MHz, and return loss RL = 15 dB. The rectangular waveguide dimensions are $a = 15.8$ mm and b = 0.63 mm, b equal to the thickness h of the employed substrate. We have used a Taconic CER-10 substrate with $\varepsilon_r = 10$ and $\tan(\delta) = 0.0035$. Irises in all cases have a thickness t = 3 mm.

The filter center frequency f_0 and bandwidth BW are expressed by:

$$f_0 = \sqrt{f_1 f_2}, \quad BW = f_1 - f_2, \tag{5}$$

which give $f_1 = 3.7$ GHz, $f_2 = 4.3$ GHz, and the filter relative bandwidth is:

$$\Delta = \frac{\lambda_{g1} - \lambda_{g2}}{\lambda_{g0}} = 0.3636 \tag{6}$$

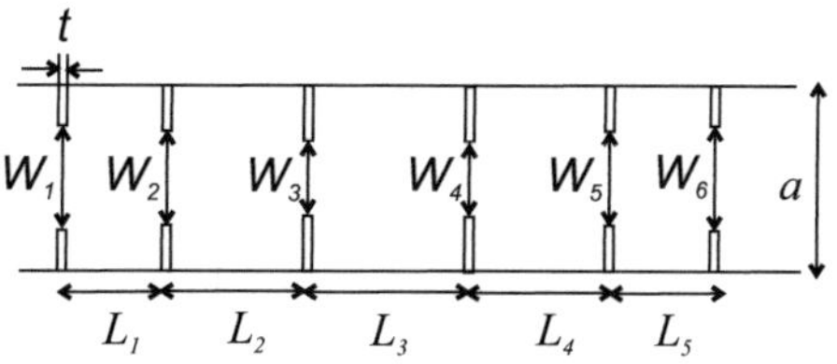

Figure 5.
Fifth order classical iris-type rectangular waveguide filter.

Then, the values obtained for the impedance inverter factors are:

$$\frac{K_{01}}{Z_0} = \frac{K_{56}}{Z_0} = \sqrt{\frac{\pi\Delta}{2g_0g_1}} = 0.7051 \tag{7}$$

$$\frac{K_{12}}{Z_0} = \frac{K_{45}}{Z_0} = \frac{\pi\Delta}{2\sqrt{g_1g_2}} = 0.45546 \tag{8}$$

$$\frac{K_{23}}{Z_0} = \frac{K_{34}}{Z_0} = \frac{\pi\Delta}{2\sqrt{g_2g_3}} = 0.34706 \tag{9}$$

Using an electromagnetic simulator, the scattering parameters of a rectangular iris (referred to the discontinuity planes) can be obtained, which are related to the T network elements shown in **Figure 4(a)**, X_s and X_p, by the following Eqs. (10) and (11):

$$j\frac{X_s}{Z_0} = \frac{1 - S_{12} + S_{11}}{1 - S_{11} + S_{12}} \tag{10}$$

$$j\frac{X_p}{Z_0} = \frac{2S_{12}}{(1 - S_{11})^2 - S_{12}^2} \tag{11}$$

where S_{11}, S_{21}, and S_{12} are the scattering parameters of the TE_{10} fundamental mode of the input waveguide at the filter center frequency f_0. For the impedance inverter shown in **Figure 4(b)**, X_s and X_p are related to K/Z_0 and φ by:

$$\frac{K}{Z_0} = \left|\tan\left(\frac{\varphi}{2}\operatorname{atan}\frac{X_s}{Z_0}\right)\right| \tag{12}$$

$$\varphi = -\operatorname{atan}\left(2\frac{X_p}{Z_0} + \frac{X_s}{Z_0}\right) - \operatorname{atan}\frac{X_s}{Z_0} \tag{13}$$

The scattering parameters of several iris of different widths have been obtained, so the values of the iris widths for this filter are $W_1 = W_6 = 12.4$ mm, $W_2 = W_5 = 10.65$ mm, and $W_3 = W_4 = 9.85$ mm. For these values of iris widths, the values of the phases provided by Eq. (13) are: $\varphi_1 = \varphi_6 = -1.7$ rad, $\varphi_2 = \varphi_5 = -1.28$ rad, and $\varphi_3 = \varphi_4 = -1.07$ rad. Finally, the resonator lengths are obtained as:

$$L_n = \frac{\lambda_{g0}}{2\pi}\left[\pi + \frac{1}{2}(\varphi_n + \varphi_{n+1})\right], \quad n = 1, \dots, N \tag{14}$$

so the values of the resonator lengths are $L_1 = L_5 = 9.4$ mm, $L_2 = L_4 = 11.2$ mm, and $L_3 = 11.8$ mm. **Figure 6** shows the simulated response of the designed rectangular waveguide filter.

The final step is to obtain the equivalent waveguide and iris widths in SIW technology with the equivalence given by Eq. (2), considering via holes diameter of $d = 0.7$ mm and separation of $s = 0.95$ mm, and also the design of microstrip to SIW transitions. For the microstrip to SIW transition, a microstrip taper has been implemented [5]. Finally, an optimization process of the designed filter response has been performed, providing the following final filter parameters: $a_{1SIW} = a_{6SIW} =$ 11.86 mm, $a_{2SIW} = a_{5SIW} = 10.48$ mm, $a_{3SIW} = a_{4SIW} = 9.90$ mm, $L_1 = L_5 = 9.4$ mm, $L_2 = L_4 = 11.2$ mm, $L_3 = 11.8$ mm, $W_t = 2.60$ mm, $L_t = 7.08$ mm, and $W_m = 0.6$ mm.

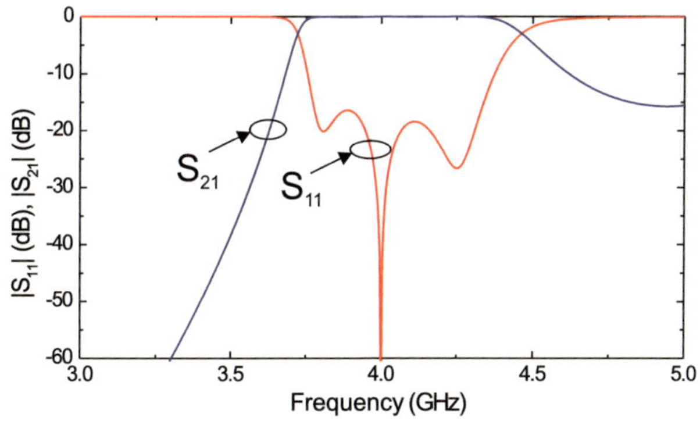

Figure 6.
Electrical response of the designed rectangular waveguide iris filter.

The designed iris SIW filter with its final dimensions is shown in **Figure 7(a)**, while its simulated and measured response is shown in **Figure 7(b)** with solid and dashed lines, respectively, showing a good impedance matching in the passband (better than 12.5 dB), and also a good out of band rejection performance (better than 20 dB).

3.2 SIW filters based on high and low dielectric constant sections

By combining the filter design procedure detailed in the previous section with the obtained results in Section 2, the same concept of band-pass filter in SIW technology

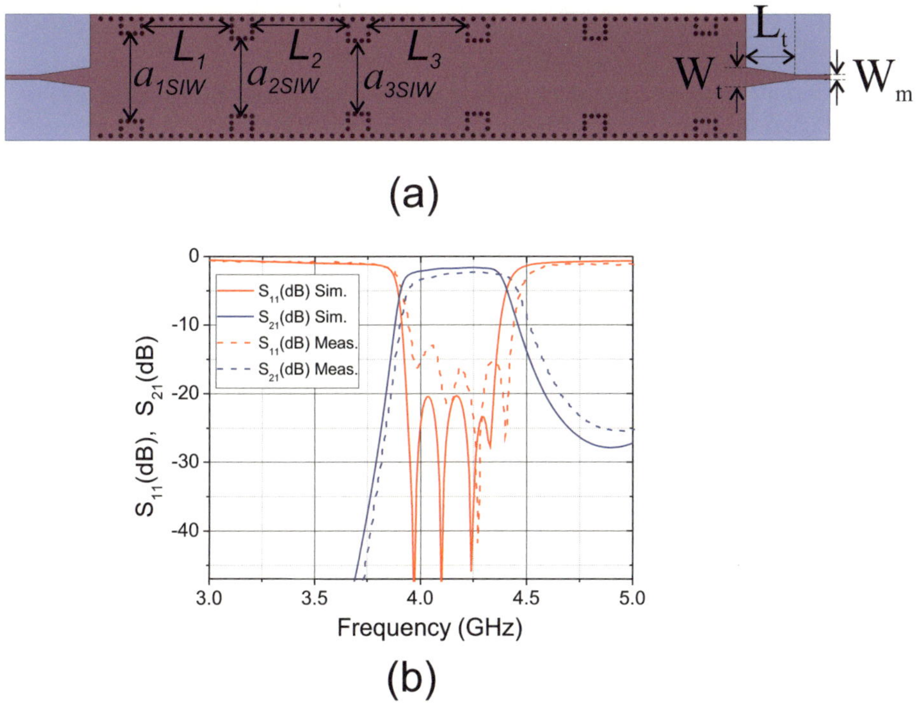

Figure 7.
(a) Scheme of the designed iris filter in SIW technology. (b) Simulated and measured response of the iris filter in SIW technology.

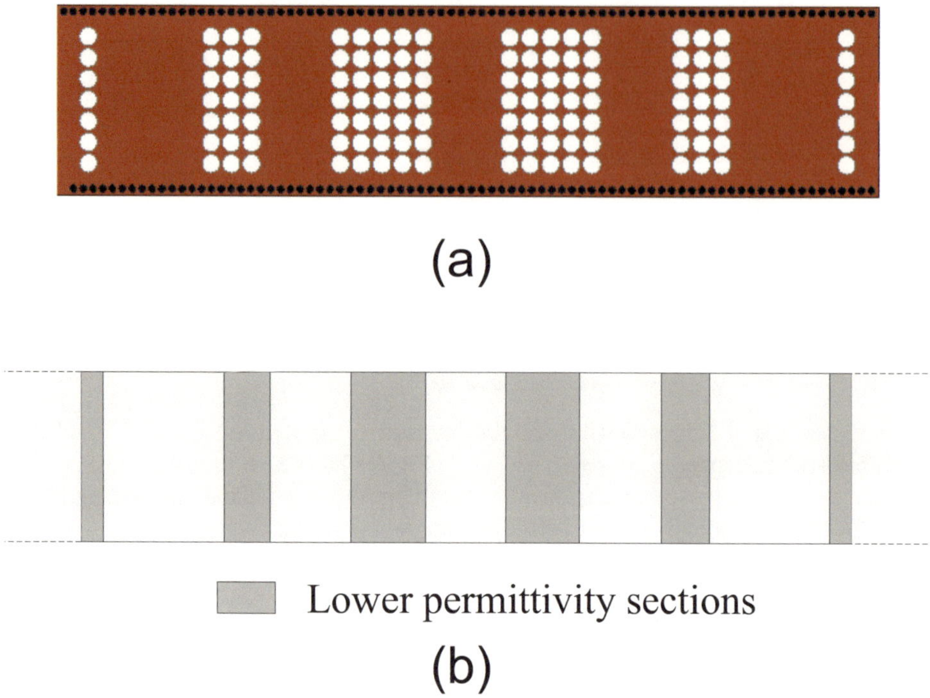

Figure 8.
SIW filter with periodic perforations. (a) Physical geometry of the filter. (b) Equivalent structure based on the homogeneous permittivity of the perforated area.

can also be implemented by exploiting SIW sections with reduced (perforated) or increased (periodically loaded with metallic cylinders) effective permittivity with ordinary SIW sections. For instance, the SIW may be perforated in some regions (see **Figure 8(a)**) to synthesize evanescent mode sections, as it has been proposed in Ref. [7], where the perforations in the dielectric substrate allow for reduction of the local effective permittivity, thus creating waveguide sections below cutoff (see **Figure 8(b)**). The lengths of the evanescent perforated waveguide sections, which are related to the impedance inverter factors in the impedance inverter model (see **Figure 8(a)**), are related to the number of hole columns. An example of SIW filter implementation with periodic perforations is shown in **Figure 9(a)** along with its simulated and measured response (**Figure 9(b)**), whose waveguide width, vias parameters, and employed substrate are the same as in Subsection 3.1. It is worth mentioning that the depth of the upper rejection band of the filter observed around 5 GHz (see **Figure 9(b)**) is directly related to the value of the reduced effective permittivity obtained in the perforated regions.

On the other hand, the combination of lower-permittivity (perforated) SIW sections with higher-permittivity (loaded with metallic cylinders) SIW sections can yield a better performance of this filter topology in terms of the rejection band, due to a higher contrast of permittivities, along with a reduction of the transversal dimension of the waveguide. An example of this phenomenon can be observed in the band-pass filter design shown in **Figure 10(a)**, constituted by the combination of rectangular perforations of the dielectric substrate and the insertion of metallic cylinders, where the SIW width has been reduced by a factor of 2 with respect to the SIW width

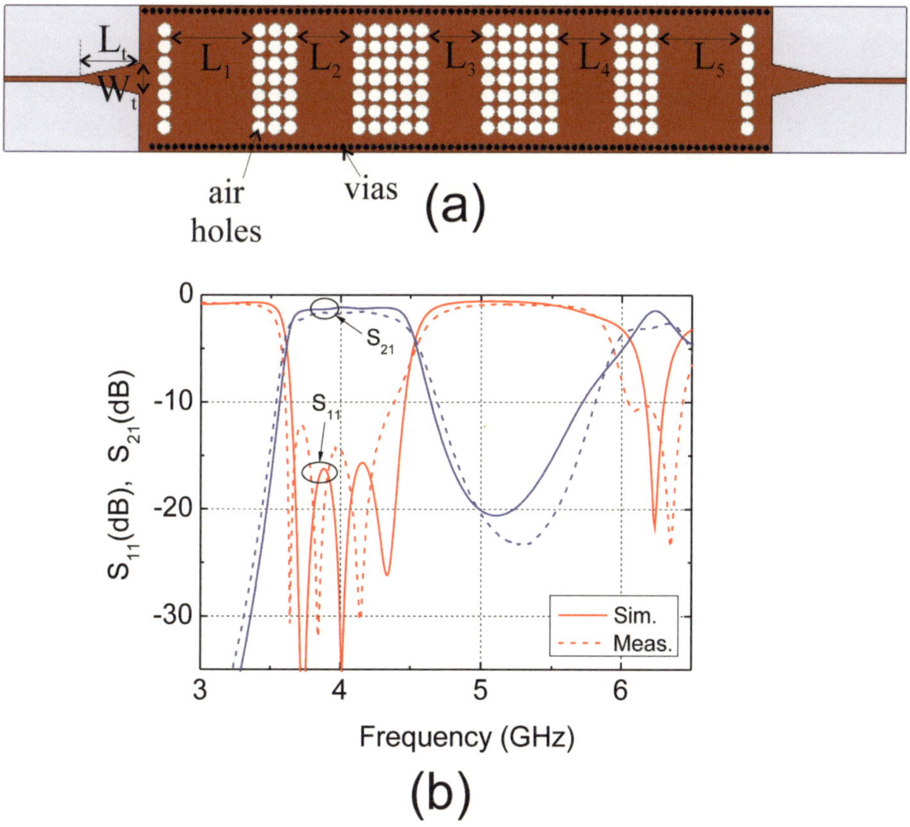

Figure 9.
(a) Example of SIW filter implementation with periodic perforations. Parameters of the air holes: $d_a = 1.7$ mm and $s_a = 1.95$ mm. Filter parameters: $L_1 = L_5 = 10.12$ mm, $L_2 = L_4 = 6.79$ mm, $L_3 = 6.42$ mm. $W_t = 3.74$ mm, and $L_t = 7.37$ mm. (b) Simulated and measured response of this filter.

employed in the filter of **Figure 9** for a similar center frequency, and with the same dielectric permittivity (being the employed substrate in the filter of **Figure 10** of thickness $b = 1.5$ mm). In this case, the perforations of the dielectric substrate have been done with a rectangular cross-section for a better selection of their widths. The simulated response of this filter is represented in **Figure 10(b)**, which reveals a deeper and wider rejection band in this case.

The perforated SIW filters described above have proven to show a good performance, exhibiting lower sensitivity to fabrication inaccuracies compared to iris-type filters with analogous frequency response. However, a limitation of such structures is that the length of the evanescent waveguide sections depends on the number of hole columns, and only discrete values are possible. To overcome it, with the aim to add flexibility to the design, a gap between the central hole rows in the perforated evanescent waveguide sections can be inserted, as it has been proposed in Ref. [12] (see **Figure 11(a)**), so a wide range of coupling coefficients can be achieved with them by changing the number of hole columns and the central gap a. This allows to design filters with the desired passband—narrow band filters, which require small couplings, can be obtained by increasing the length of the waveguide sections below the cutoff (i.e., the number of hole columns), and reducing the central gaps, and vice versa. An example of

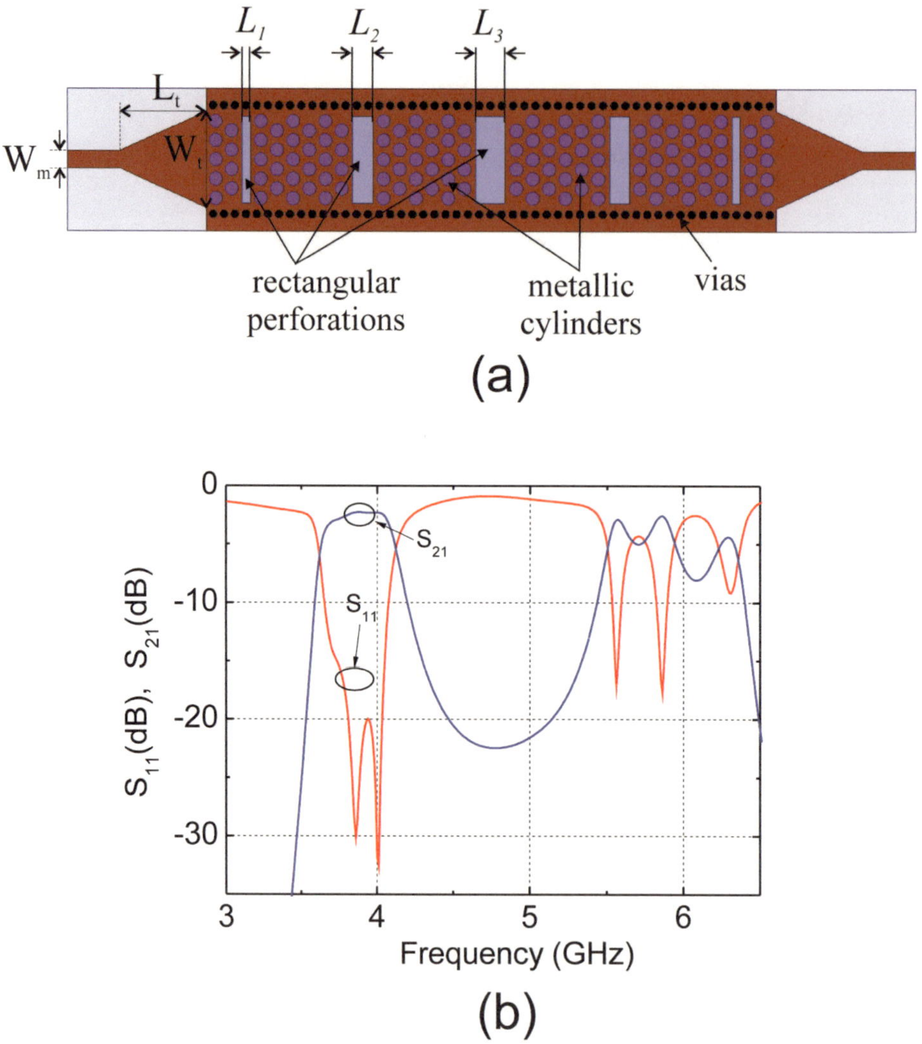

Figure 10.
(a) Example of SIW filter implementation with the combination of rectangular perforations of the dielectric substrate and the insertion of metallic cylinders. a_{SIW} *= 9.4 mm and b = 1.5 mm. Parameters of the metallic cylinders:* d_c*= 1.1 mm,* s_c *= 1.6 mm, and thickness of 1.25 mm. Parameters of the rectangular perforations:* $L_1 = L_5 = 0.7$ *mm,* $L_2 = L_4 = 1.7$ *mm,* $L_3 = 2.5$ *mm.* $W_t = 8.05$ *mm,* $L_t = 8.1$ *mm, and* $W_m = 1.45$ *mm. (b) Simulated response of this filter.*

SIW filter implementation employing Taconic CER-10 substrate with these guidelines can be seen in **Figure 11**, with the geometrical dimensions of the filter and a photograph of the fabricated prototype, along with its simulated and measured response.

Finally, further developments of SIW filters with perforations of the dielectric substrate extending this concept to half-mode SIW structures can be done [12], with the aim to reduce the size of the filter. As an example, the perforated SIW filter in **Figure 11** has been applied to the half-mode SIW configuration by removing half of the top metal layer (where an HFSS reoptimization has been done). **Figure 12** shows the geometry of the filter with the geometrical dimensions and a photograph of the top layer, along with the comparison of the simulated and measured response. In this

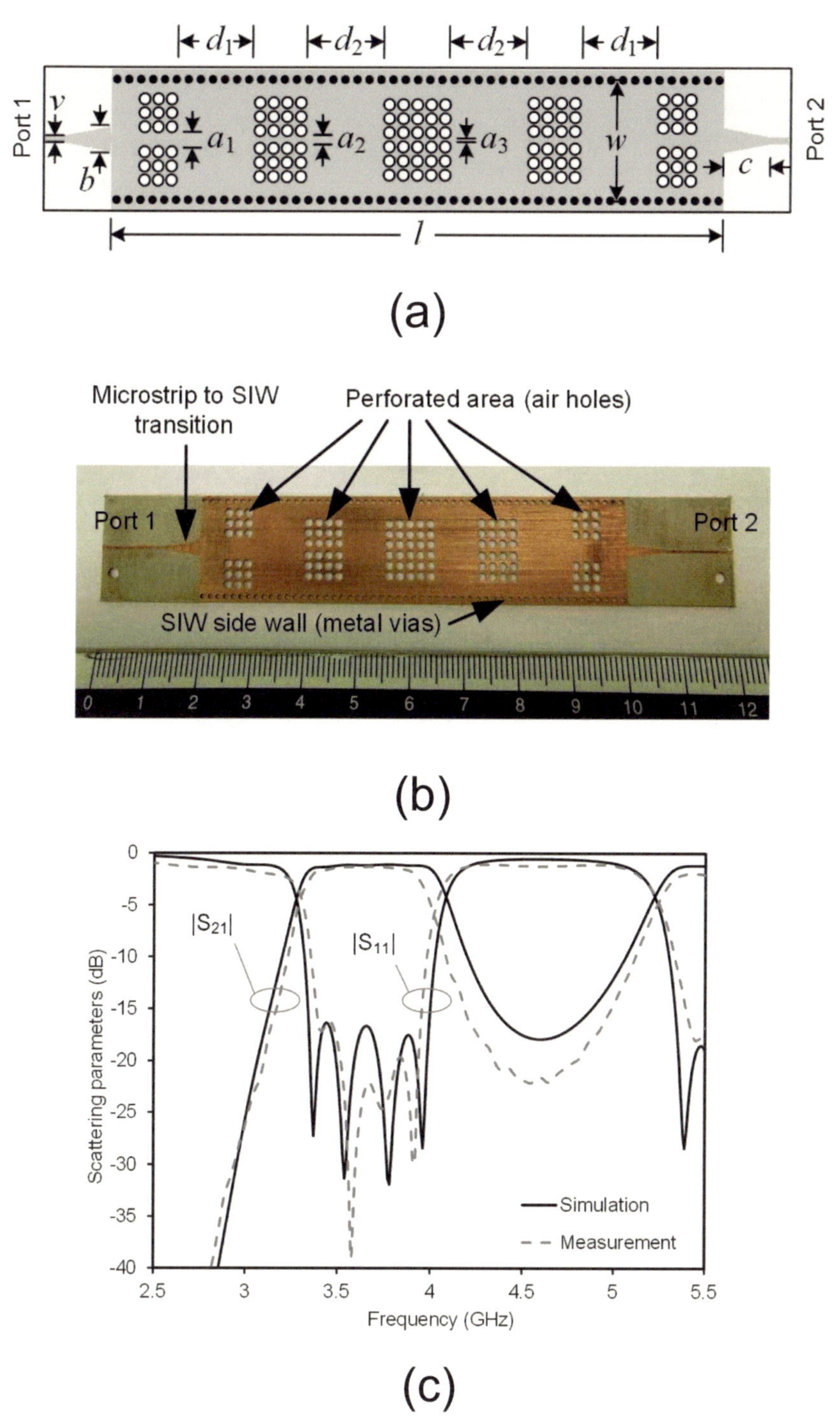

Figure 11.
Example of a four-pole perforated SIW filter incorporating gaps between the central hole rows in the perforated evanescent waveguide sections. (a) Geometry of the filter (dimensions in millimeter: v = 0.6, b = 2.6, d_1 = 10, d_2 = 7.95, a_1 = 4, a_2 = 0.55, a_3 = 0.25, w = 17.8, c = 7, and l = 81). (b) Photograph of the prototype. (c) Scattering parameters of the four-pole filter (HFSS simulation compared with measured data). Reprinted with permission from Ref. [12]; copyright 2017 IEEE.

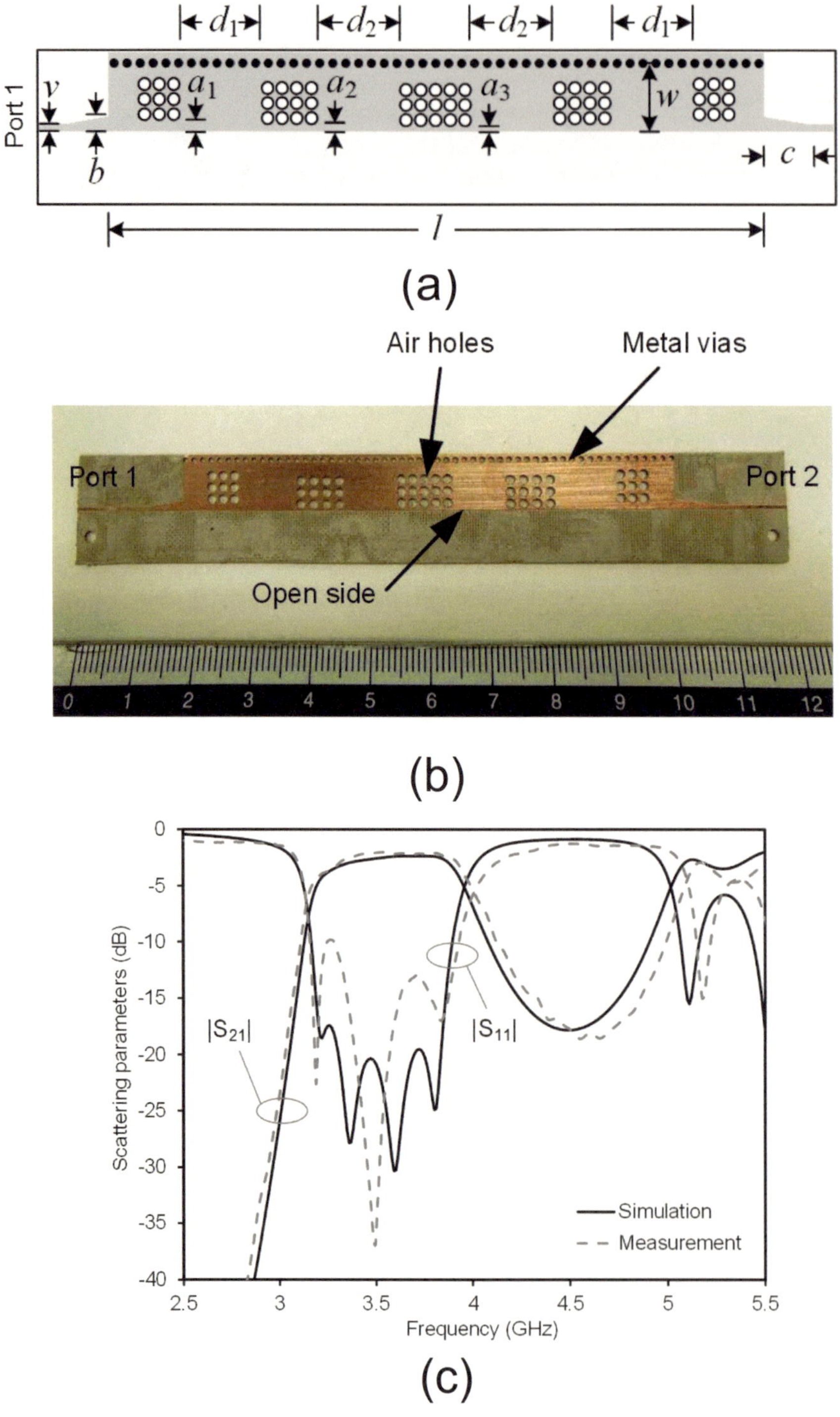

Figure 12.
Example of a four-pole filter based on perforated half-mode SIW structure. (a) Geometry of the filter (dimensions in millimeter: $v = 0.6$, $b = 1.6$, $d_1 = 9.6$, $d_2 = 8.35$, $a_1 = 1$, $a_2 = 0.2$, $a_3 = 0.4$, $w = 8.3$, $c = 7$, and $l = 83$). (b) Photograph of the prototype. (c) Electrical response of the half-mode filter (HFSS simulation compared with measured data). Reprinted with permission from Ref. [12]; copyright 2017 IEEE.

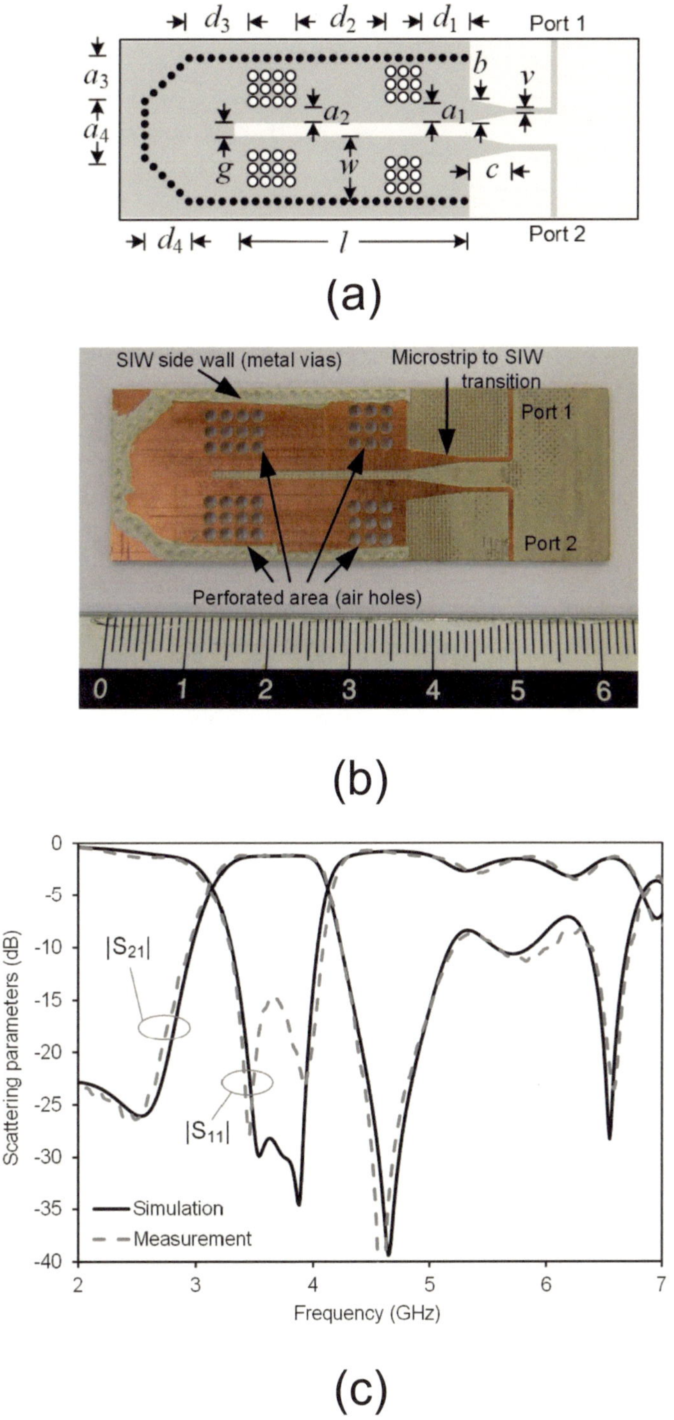

Figure 13.
Example of a three-pole half-mode SIW filter in folded configuration. (a) Geometry of the filter (dimensions in millimeter: v = 0.6, b = 2.6, d_1 = 1.65, d_2 = 10.56, d_3 = 3.9, d_4 = 6.7, a_1 = 2.45, a_2 = 1.95, a_3 = 5, a_4 = 9.45, w = 9.2, c = 7, g = 1, and l = 25.1). (b) Photograph of the prototype. (c) Electrical response of the folded filter (HFSS simulation compared with measured data). Reprinted with permission from Ref. [12]; copyright 2017 IEEE.

	Number of resonators	Center freq. (GHz)	FBW* (%)	IL (dB)	Minimum RL (dB)	Size (mm^2)
[13]	2	4.75	2.5	1.4	13	1085
[14]	1	7.4	16.9	2.0	18	66.6
[15]	2	10.0	0.02	1.0	20	401
This work	5	4.0	15	1.6	12	327.4

**FBW: fractional bandwidth.*

Table 1.
*Comparison of performance of the SIW filter topology of this work (filter shown in **Figure 9**) with other filter topologies.*

case, a significant size reduction of the circuit has been achieved, although the structure is affected by radiation leakage (which is directly related to the observed higher insertion loss of the filter), due to the field distribution along the open boundary of the half-mode SIW. A folded filter configuration can be adopted to mitigate radiation losses of the half-mode SIW filter (see **Figure 13**), so the open boundaries of the half-mode SIW structure are located face-to-face, which reduces the radiation loss, and at the same time introduces a transmission zero in the frequency response through the direct input–output coupling. As an example, **Figure 13(a)** shows a three-pole half-mode SIW filter in folded configuration (geometry of the filter and final dimensions), while a photograph of the filter and its electrical response is shown in **Figure 13(b)** and **(c)**. The effect of the absence of radiation leakage is the flat insertion loss observed in **Figure 13(c)**.

A comparison between the obtained results of one of the filters in SIW technology presented in this chapter is based on stepped-impedance configurations with high and low dielectric constant sections (filter shown in **Figure 9**), and some other band-pass filters in SIW technology reported in the technical literature are presented in **Table 1**. As can be seen from **Table 1**, the proposed filters in this work show a clear improvement in bandwidth and insertion losses with respect to similar band-pass SIW filter topologies reported in the technical literature.

4. Conclusions

In this chapter, we provide guidelines for the design of novel SIW filters based on the alternation of SIW sections with reduced (perforated) or increased (periodically loaded with metallic cylinders) effective permittivity with ordinary SIW sections, making use of the impedance inverter model. Examples of several SIW filter configurations show that this solution can lead to the design of compact filters with good performance and low sensitivity to fabrication inaccuracies.

Acknowledgements

A. Coves acknowledges grant PID2019-103982RB-C43 funded by Ministerio de Ciencia e Innovación (MCIN)/Agencia Estatal de Investigación (AEI)/10.13039/501100011033.

Author details

Angela Coves[1*] and Maurizio Bozzi[2]

1 University Miguel Hernández de Elche-I3E, Elche, Spain

2 University of Pavia, Pavia, Italy

*Address all correspondence to: angela.coves@umh.es

IntechOpen

© 2022 The Author(s). Licensee IntechOpen. This chapter is distributed under the terms of the Creative Commons Attribution License (http://creativecommons.org/licenses/by/3.0), which permits unrestricted use, distribution, and reproduction in any medium, provided the original work is properly cited.

References

[1] Bozzi M, Georgiadis A, Wu K. Review of substrate integrated waveguide (SIW) circuits and antennas. IET Microwaves, Antennas and Propagation. 2011;**5**: 909-920. DOI: 10.1049/iet-map. 2010.0463

[2] Wu K, Bozzi M, Fonseca JG. Substrate integrated transmission lines: Review and applications. IEEE Journal of Microwaves. 2021;**1**:345-363. DOI: 10.1109/JMW.2020.3034379

[3] Cassivi Y, Perregrini L, Arcioni P, Bressan M, Wu K, Conciauro G. Dispersion characteristics of substrate integrated rectangular waveguide. IEEE Microwave and Wireless Components Letters. 2002;**12**:333-335. DOI: 10.1109/ LMWC.2002.803188

[4] Xu F, Wu K. Guided-wave and leakage characteristics of substrate integrated waveguide. IEEE Transactions on Microwave Theory and Techniques. 2005;**53**:66-73. DOI: 10.1109/TMTT.2004.839303

[5] Deslandes D, Wu K. Integrated microstrip and rectangular waveguide in planar form. IEEE Microwave and Wireless Components Letters. 2001;**11**:68-70. DOI: 10.1109/ 7260.914305

[6] ANSYS High Frequency Structure Simulator (HFSS). Available from: https://www.ansys.com/products/ electronics/ansys-hfss. [Accessed: February 19, 2022]

[7] Coves A, Torregrosa-Penalva G, San-Blas AA, Sánchez-Soriano MA, Martellosio A, Bronchalo E, et al. A novel band-pass filter based on a periodically drilled SIW structure. Radio Science. 2016;**51**:328-336. DOI: 10.1002/ 2015RS005874

[8] Nguyen HV, Gauthier J, Fernández JM, Sierra-Castañer M, Caloz C. Metallic Wire Substrate (MWS) for miniaturization in planar microwave applications. In: Proceedings of Asia-Pacific Microwave Conference; Yokohama, Japan; December 2006. pp. 1084-1087

[9] Njoku CC, Whittow WG, Vardaxoglou JC. Simulation methodology for synthesis of antenna substrates with microscale inclusions. IEEE Transactions on Antennas and Propagation. 2012;**60**:2194-2202. DOI: 10.1109/TAP.2012.2189736

[10] Vicent G, Coves A, Bronchalo E, Torregrosa G. Synthesis of an rtificial high effective permittivity medium in a SIW periodically loaded with metallic cylinders. In: 2017 Progress in Electromagnetics Research Symposium (PIERS); 22–25 May 2017; St. Petersburg, Russia. pp. 3573-3578

[11] Cameron RJ, Kudsia CM, Mansour RR. Microwave Filters for Communication Systems: Fundamentals, Design, and Applications. 2nd ed. Hoboken, New Jersey, U.S.: John Wiley & Sons, Inc; 2018. DOI: 10.1002/ 9781119292371

[12] Silvestri L, Massoni E, Tomassoni C, Coves A, Bozzi M, Perregrini L. Substrate integrated waveguide filters based on a dielectric layer with periodic perforations. IEEE Transactions on Microwave Theory and Techniques. 2017;**65**:2687-2697. DOI: 10.1109/ TMTT.2004.839303

[13] Silvestri L, Ghiotto A, Tomassoni C, Bozzi M, Perregrini L. Partially air-filled substrate integrated waveguide filters with full control of transmission zeros. IEEE Transactions on Microwave Theory

and Techniques. 2019;**67**:3673-3682. DOI: 10.1109/TMTT.2019.2926356

[14] Jin B, Zhang P, Mu J, Zhang M, Li M. A miniaturized bandpass filter basing on HMSIW loaded dual-mode CSRR. In: Proceedings of the International Wireless Symposium (IWS '21); 23-26 May 2021; Nanjing. China: IEEE; 2021. pp. 866-870. DOI: 10.1109/IWS52775.2021.9499651

[15] Zhu F, Luo GQ, You B, Zhang XH, Wu K. Planar dual-mode bandpass filters using perturbed substrate-integrated waveguide rectangular cavities. IEEE Transactions on Microwave Theory and Techniques. 2021;**69**:3048-3057. DOI: 10.1109/TMTT.2021.3074617

Chapter 4

Ridge Gap Waveguide Beamforming Components and Antennas for Millimeter-Wave Applications

Mohammad Ali AbdElraheem, Mohamed Mamdouh M. Ali, Islam Afifi and Abdel R. Sebak

Abstract

With the improvement of mobile communication technologies and their broad applications, mobile communication will have more impact on our life. Such systems will support a variety of personal communication services with high-data rate and very low latency applications. To achieve such demands, many proposals associated with the development of 5G identify a set of requirements for which different technological directions are independently emerging. One direction is utilizing the millimeter-wave (mm-Wave) frequency bands where more spectrums are available. Millimeter-wave frequencies offer the advantage of physically smaller components that results in cost-effective RF transceivers and feasible large-scale integrated phased arrays. The smart RF transceivers of 5G along with the potential high-frequency innovative designs must satisfy the growing consumer and technology requirements. This implies utilizing the state-of-the-art guiding structures, especially printed ridge gap waveguide (PRGW), that have low loss and minimal dispersion compared with traditional PCB-based structures. The present chapter focuses on the necessary components for a beamforming antenna system which is implemented using PRGW technology. Millimeter wave antennas with different polarizations have been addressed. Power combining and dividing components have been also developed. These components have been used for integration in a complete beamforming antenna system working at an mm-Wave frequency band.

Keywords: millimeter wave, electromagnetic band gap (EBG) structure, printed ridge gap waveguide (PRGW), butler matrix, beam switching

1. Introduction

Wireless technology and devices are fundamental components in many aspects of life including personal communications, internet activities, sensing, and imaging applications for industrial and medical purposes. Wireless technology development is always encouraged by the needs of these applications to enhance their performance in terms of service quality and cost. A fundamental issue in wireless technology is the system operating frequency which varies according to the target application. An overview of the operating standards in wireless technology reveals that most systems nowadays operate in the microwave frequency band below 3 GHz [1–7]. The current state of wireless applications reflects the demand for using higher frequency bands to enhance the performance of the wireless system. For instance, communication technology is rapidly pushed toward next-generation networks where massive data rates, very low latency, and a high level of service integration are required to enhance the performance of the wireless communication system [8, 9]. To achieve these demands, much higher bandwidths must be used to increase the capacity of the communication channel. Imaging systems are also in the demand of using higher frequencies to overcome the resolution limits at the lower microwave band. Millimeter-Wave frequency range is beyond 30 GHz, where the relative bandwidth is equivalent to multiples of frequency channels at the sub 3 GHz range. Moreover, the signal wavelength at mm-Wave is significantly small and enables high-resolution imaging. These advantages have encouraged the move to the mm-Wave band and have ignited the spark of innovating components with superior characteristics in these high frequencies [10–13].

Many challenges are addressed in the literature, where the small-signal wavelength at the mm-Wave band results in components with small physical dimensions that need high tolerance fabrication facilities with an extremely large cost. In addition, being an advantage that enables integrating large systems in a small area, this is also a limiting property for the power that can be handled by an mm-Wave communications device [14–16]. Hence, antenna arrays must be used in mm-Wave transceivers to provide a suitable amount of gain, especially to compensate for the large path loss caused by atmospheric attenuation at frequencies like 60 GHz and such [17, 18]. Furthermore, the high level of versatility in the next generation network requires running multiple services simultaneously, a property that forces the use of a diversity technique to enable various communications without the need to add bandwidth [19, 20]. Such demand can be realized using multiple-input multiple-output (MIMO) systems employing beamforming techniques to benefit from the deployed high gain array [21, 22]. Therefore, the beamforming antenna array is an essential subject in mm-Wave research in the context of next-generation networks and imaging systems. Several techniques are proposed to implement a beamforming antenna array, which required the usage of various microwave components including power dividers, crossovers, phase shifters, hybrid couplers, and antennas [23–26]. However, the realization of these components at mm-Wave frequencies using traditional printed guiding technologies is another key challenge that needs to be tackled. Although traditional guiding structures such as microstrip line and stripline support a Q-TEM mode, which is subject to minimal dispersion, it has high radiation and dielectric losses at mm-Wave frequencies [27]. On the other hand, a modern guiding structure such as substrate integrated waveguide (SIW) is developed at mm-Wave frequencies as it has low radiation losses compared to microstrip line and stripline structures [28, 29]. However, it can support only TE mode which is subject to large dispersion and causes signal distortion [30, 31]. In addition, the signal is totally propagating inside a dielectric, which leads to high losses at mm-Wave frequencies. Therefore, a novel technology of guiding structures is introduced to provide a solution for the mentioned challenges. This technology is the printed ridge gap waveguide (PRGW) technology that was introduced as a novel low loss Q-TEM guiding structure for the mm-wave frequency range [32–40].

The printed ridge gap structure is shown in **Figure 1a**, where the operating mechanism is based on the idea of wave suppression between a perfect electric conductor

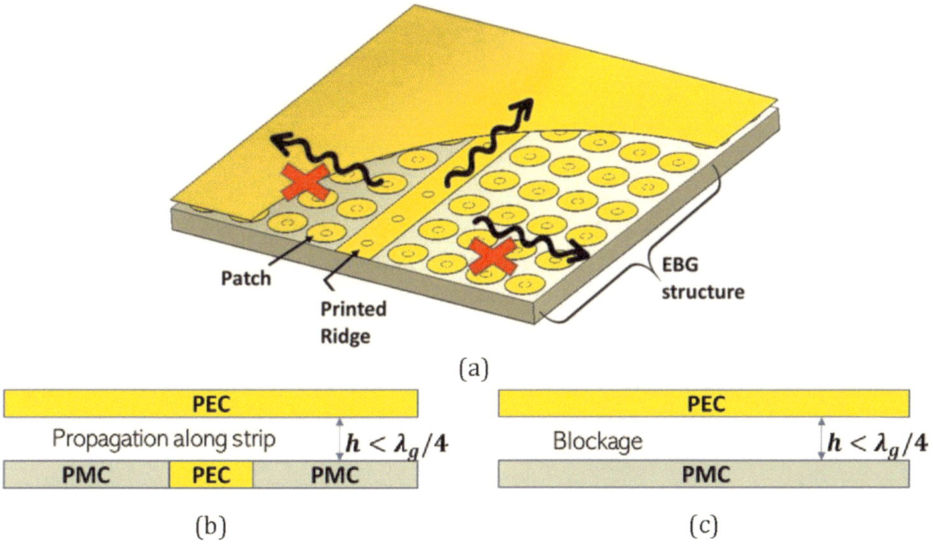

Figure 1.
Parallel Plate PEC/PMC structure. (a) PRGW with mushroom EBG cells. (b) achieving propagation with added longitudinal middle strip (c) blockage condition.

(PEC) and a perfect magnetic conductor (PMC) parallel to each other. Such parallel plate structure does not allow wave propagation unless the separation between the plates is enough for achieving zero tangential fields at one plate and zero normal at the other. This cannot happen for separations less than a quarter wavelength. Such condition is used to prevent wave leakage through the sides of the guiding structure, where adding a middle ridge allows having a propagating mode in the longitudinal direction as shown in **Figure 1b**. Since the PMC material does not exist in practice, an emulation of such material is the artificial magnetic conductors (AMCs) that can be implemented using the mushroom-like Electromagnetic Band Gap (EBG) structure in **Figure 1a**. The design of PRGW has been well addressed in the literature, which starts by designing the EBG-cells to support the required bandgap over the operating

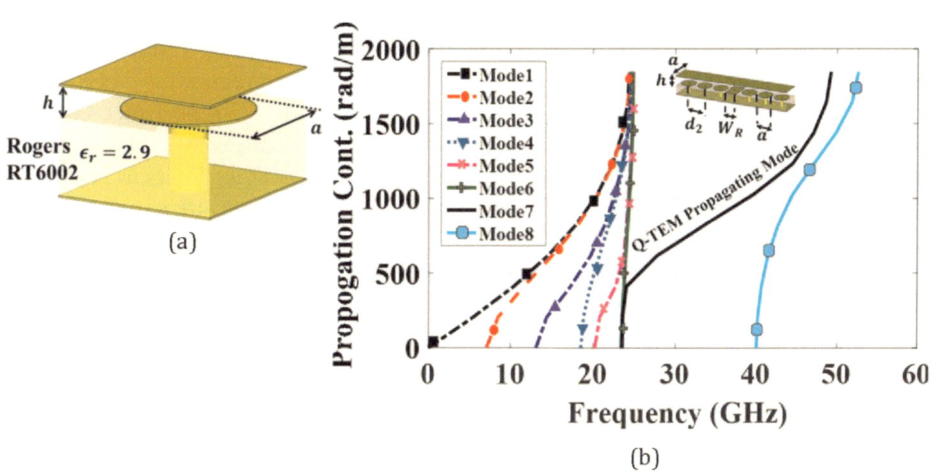

Figure 2.
Design of electromagnetic band gap (EBG) structure. (a) EBG unit cell. (b) PRGW line segment and simulated dispersion diagram showing the propagating Q-TEM mode.

Dimension	a	h	h_s	d_2	W_R
Value (mm)	1.32	0.289	0.3065	1.52	1.37

Table 1.
dimensions of unit cell in EBG structure designed for 26–40 GHz band.

bandwidth. **Figure 2a** shows the EBG unit cell used to design the PRGW feeding line, where the geometrical parameters are listed in **Table 1**. As shown in **Figure 2b**, modelling the cell on CST MWS simulation software proves to achieve a bandgap covering the entire mm-Wave Ka-band of 26–40 GHz [41].

This chapter is organized in four sections as follows: Section 2 introduces two main transitions, from microstrip line and from coaxial line to PRGW that allow the integration with other technologies. Section 3 focuses on the design of PRGW hybrid couplers, crossovers, and phase shifters, which are the main building blocks of the beam switching networks. Section 4 discusses several designs of antenna elements and arrays with various polarizations and excitation techniques. The integration of the previous components to form a beam scanning antenna system will be discussed in Section 5, while the last section concludes the introduced material.

2. Transitions

To ensure the full integrability of the PRGW technology with other TEM guiding structures such as microstrip and coaxial lines, several types of PRGW transitions have been proposed in the literature correspondingly [42, 43]. The microstrip line to PRGW is considered the most simple and straightforward transition that can be used to excite the PRGW with a deep matching level over a broad bandwidth. **Figure 3** shows the geometrical configuration of microstrip line to PRGW transition, where the PRGW is directly connected with the 50 Ω microstrip line through a taper transformer with length L_t and width W_t [44, 45]. These two parameters are then optimized to adjust the matching level over the operating bandwidth, where the optimum dimensions are listed in **Table 2**. The operation of the transition is assessed using a PRGW bend junction as shown in **Figure 3a**. Comparison between the simulated and measured S-parameter results are shown in **Figure 3c**, where a matching level below −15 dB over the operating frequency bandwidth is achieved. It can be noticed that the measured insertion loss of this type of transition reaches −1 dB, where a large part of these losses results from the microstrip line radiation losses. This results in an inaccurate assessment of the PRGW devices through using this type of transition [44, 45].

Therefore, another type of transition, from coaxial to PRGW, is proposed to reduce the radiation losses, with the configuration illustrated in **Figure 4**. Such transition is useful in many circumstances where the PRGW device is the first component in the system taking the feed from the output coaxial terminal of the source. Like the previous design, mushroom shape periodic patch cells are used for emulating the artificial magnetic conductor. Two substrate materials are used in this multilayer configuration with an empty region in one substrate to provide the air gap [46]. Additionally, a group of metal vias is drilled around the transition to the ground plane to enhance the device performance [46]. The design dimensions are optimized to cover the whole band of 24–42 GHz and are listed in **Table 3**, while the simulation and measurement results in **Figure 4c** assess that operation.

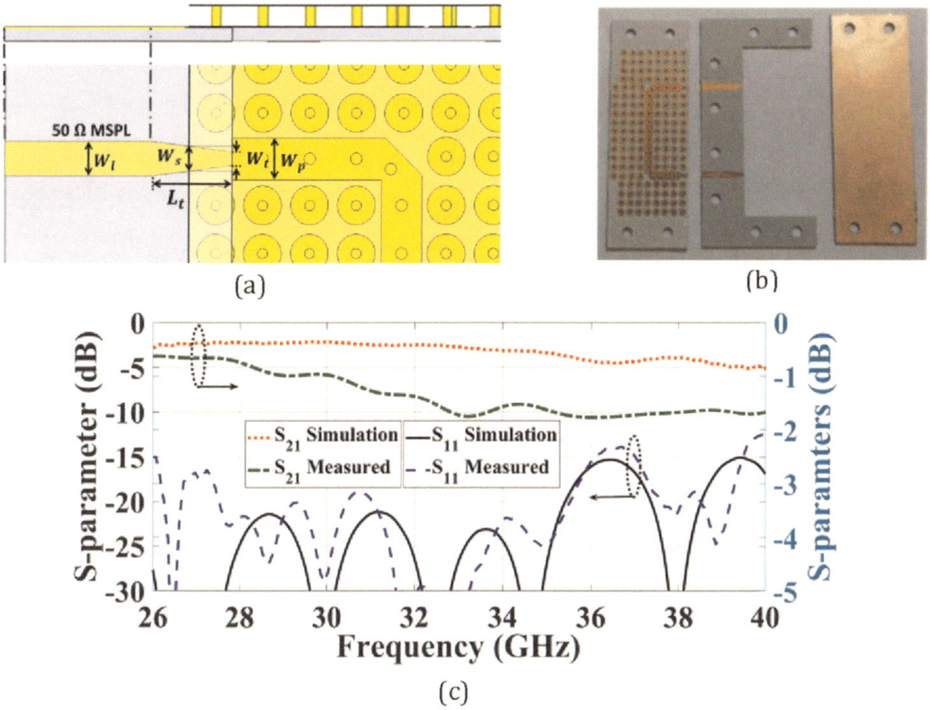

Figure 3.
Microstrip to printed ridge gap waveguide transition. (a) Front and side views of the model. (b) Fabricated transition. (c) simulated and measured S-parameter responses.

Dimension	L_t	W_t	W_s	W_l	W_p
Value (mm)	3.045	0.5	0.946	1.265	1.55

Table 2.
Dimensions of the microstrip\PRGW transition illustrated in ***Figure 3****.*

3. Feeding structures

3.1 Hybrid couplers

Hybrid couplers are the main building blocks of the beam switching network, which is used to divide the power equally with 90° phase shift and high isolation between the ports [47]. The design and analysis of several PRGW hybrid couplers configurations have been proposed in the literature, where two featured designs will be discussed in this section as are mainly deployed in beam switching antenna systems [45, 48–50].

The first design is shown in **Figure 5**, where four identical PRGW lines are connected through a rectangular coupling section with dimensions ($L \times W$). These dimensions mainly control the coupling in the desired operating bandwidth, where the initial dimensions are calculated through applying the even/odd mode analysis [49]. Since the impedance of the four PRGW ports is different from the impedance of the coupling section, a taper matching transformer is introduced to achieve a deep matching level over the operating frequency band, where the final dimensions of the

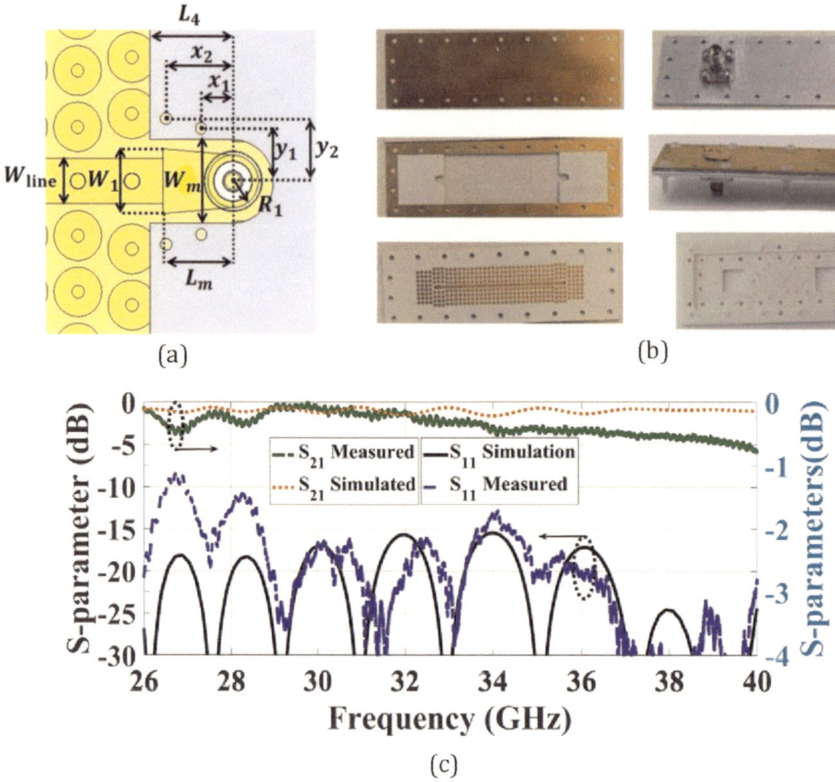

Figure 4.
The design of coaxial to PRGW transition. (a) transition section with detailed dimensions. (b) fabricated layers. (c) simulated and measured results of the scattering parameters.

Dimension	R_1	L_m	W_1	W_{line}	a	d_{via}	d_{cap}	h_{pin}
Value (mm)	0.79	2.2	1.95	1.38	1.7	0.39	1.5	2.8
Dimension	L_4	W_2	d_{via1}	L_{line}	x_1	y_1	x_2	y_2
Value (mm)	2.6	2.508	0.35	52.2	1	1.6	1.1	1.9

Table 3.
Dimensions of the coaxial to PRGW transition in ***Figure 4****.*

coupler are listed also in **Table 4**. The performance of the coupler is evaluated through simulation, where −15 dB matching level and isolation over a relative bandwidth of 26.5% at 30GHz are achieved as shown in **Figure 5b**. In addition, $90° \pm 5°$ phase shift is achieved between the output ports over the whole operating bandwidth [49]. However, one major drawback of this coupler is the amplitude imbalance (3.5 dB ± 1.5) is large, where high performance beam switching systems require both precise amplitude and phase balance over the operating frequency band.

Therefore, an alternative model targeting the same frequency band is presented in **Figure 6a**, where a circular coupling section is used rather than the rectangular junction in the former design [51]. The circular junction consists of two rings with different radii representing the widths of the equivalent branched line directional coupler. A bowtie shape slot is introduced in the center of the rings with specific

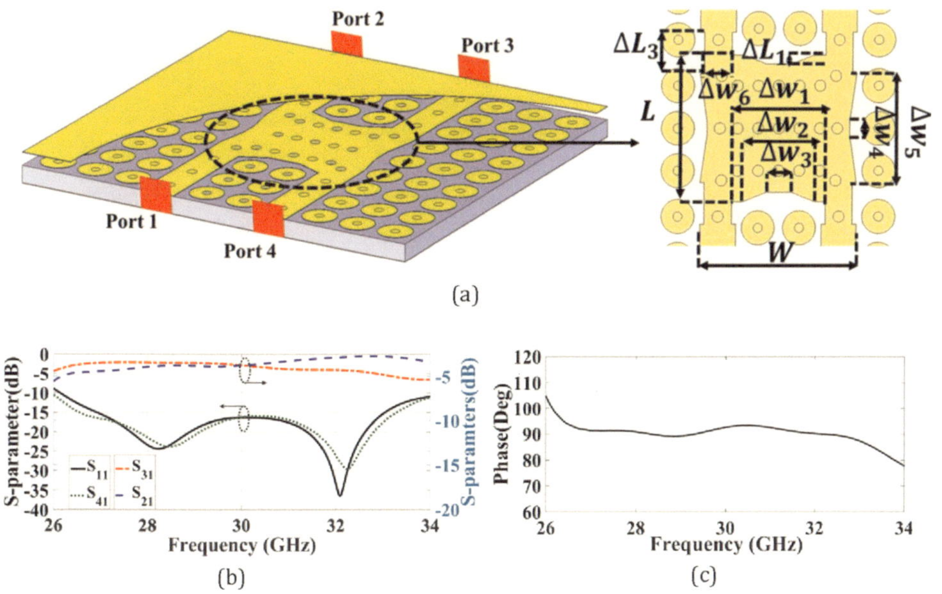

Figure 5.
Quadrature 3 dB hybrid coupler: (a) design details and dimensions. (b) S-parameters simulation results. (c) phase difference of output ports.

Parameter	W	L	ΔL_1	ΔL_3	Δw_1	Δw_2	Δw_3	Δw_4	Δw_5	Δw_6
Value (mm)	6.7	6.1	0.32	2.1	4.1	3	1.1	0.7	4.5	0.9

Table 4.
Dimensions of the coupling section illustrated in ***Figure 5a****.*

orientation to adjust the coupler performance [51]. Furthermore, a step matching section is added to each transmission line to adapt the impedances over the operating band. The design depends on the suitable adjustment of each ring width, the bowtie slot, and the dimensions of the matching section. These parameters are given in **Table 5**, while the simulation results are illustrated in **Figure 6b** and **c** proving efficient operation over the frequency range of 26.4–33.75 GHz. As an assessment of the advantage of PRGW technology over other new technologies, **Table 6** summarizes a comparison among the performance of the rectangular hybrid coupler in **Figure 5** and other designs in the literature.

3.2 Crossover

Beam switching networks impose the usage of crossover connections when two PRGW lines cross each other at a point while at the same time must be totally isolated [55, 56]. Two main techniques can be used to implement crossovers, and two featured designs will be presented in this section [51, 57].

The first design is based on the traditional technique of cascading 3 dB directional couplers to implement the crossover. However, three quarter wavelength sections are used to widen the bandwidth more than that of the traditional designs which use two sections only [51]. Using three cascaded sections introduces more design variables and more degrees of freedom to optimize the performance. The analysis of the structure is

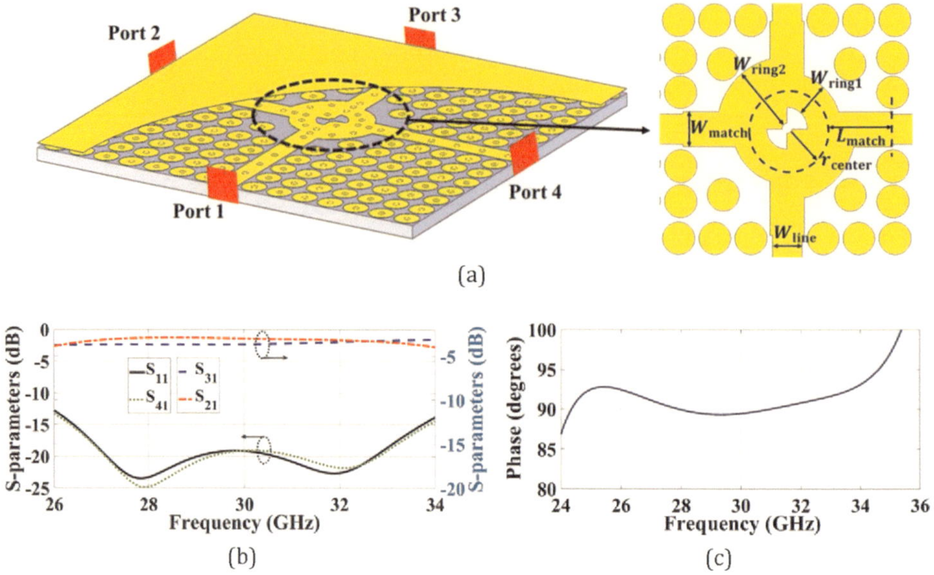

Figure 6.
Design and performance of the ring coupler. (a) design details and dimensions (b) full S-parameters for a single port. (c) phase difference between output ports.

Parameter	W_{line}	W_{match}	L_{match}	r_{center}	W_{ring1}	W_{ring2}
Value (mm)	1.38	1.582	3.14	1.753	1.595	3.05

Table 5.
Dimensions of the design in ***Figure 6a****.*

done by even\odd mode analysis, and the design parameters are optimized for minimum isolation and reflection [58]. The design is shown in **Figure 7a** and **b**, while the parameters are given in **Table** 7. As seen by the results in **Figure** 7**c**, the device achieves a relative bandwidth of 23% at 30 GHz with more than 15 dB isolation.

A disadvantage of the traditional technique of using cascaded couplers is the large size of the crossover. Aiming to avoid that common shortage, another model is presented in **Figure 8a**. This design is based on achieving 0 dB coupling in a directional coupler by designing even and odd impedances with 5% difference between each other over the frequency band of interest [57]. The design has the structure of four PRGW lines connected through a rectangular coupling section in the middle. By suitable choice of the dimensions of the coupling section, full isolation can be ensured between a single port and two of the four ports, yielding 0 dB coupling with the remaining port [57]. Multiple steps are added in the coupling section as tuning parameters to enhance the operating bandwidth [57]. The final dimensions of these steps and the coupling section are given in **Table 8**, for which the device produces the S-parameters illustrated by **Figure 8c** and obtained through CST simulation. These results indicate acceptable operation over the frequency range of 28.5–32.5 GHz in terms of isolation and coupling. The device achieves relative bandwidth of about 13.3%, which is typically higher than usual single layer crossover designs of the same size. Moreover, comparison with designs from other technologies is listed in **Table 9**, featuring the advantages of both the compact size and high relative bandwidth obtained by PRGW technology.

Reference	Technology	Impedance bandwidth	Amplitudebalance (dB)	Phase balance	Size ($\lambda_0 \times \lambda_0$)
[52]	Rectangular waveguide	6.5% at 14 GHz	3.1 ± 0.2 BW = 6.5%	$90^0 \pm 0.3^0$ BW = 6.5%	1.6×1.7
[53]	Substrate integrated waveguide (SIW)	18% at 24 GHz	4.7 ± 0.5 BW = 10%	$92^0 \pm 2^0$ BW = 18%	1.4×1.5
[54]	Half mode SIW	11% at 27.5 GHz	4.8 ± 0.25 BW = 7%	$180^0 \pm 10^0$ BW = 11%	Not included
	Proposed PRGW	26% at 30 GHz	3.7 ± 0.75 BW = 13%	$90^0 \pm 5^0$ BW = 23%	1.3×1.1

Table 6.
Comparison of PRGW hybrid coupler design with other technologies.

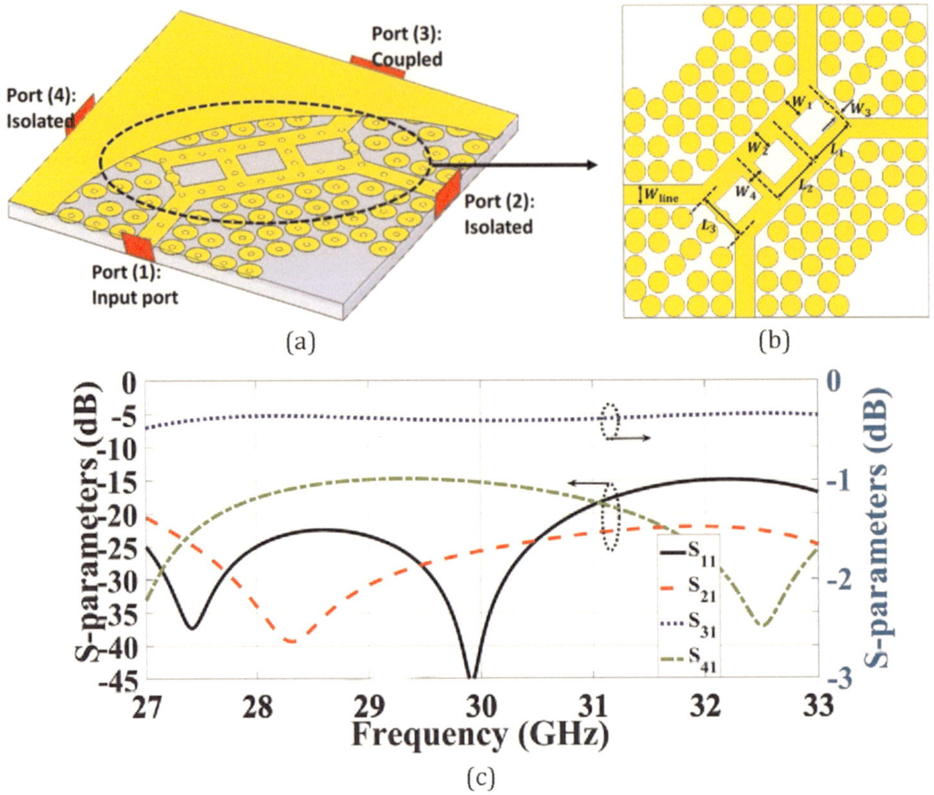

Figure 7.
Crossover design by cascaded couplers. (a)3D view of the PRGW crossover. (b) details and dimensions of the cascaded couplers. (c) simulated S-parameter response of the crossover along isolation and coupling directions.

Parameter	W_{line}	W_1	W_2	W_3	W_4	L_1	L_2	L_3
Value (mm)	1.38	1.56	1.49	0.52	1.23	3.65	3.65	3.65

Table 7.
Parameters of the crossover illustrated by ***Figure 7b****.*

3.3 Phase shifters

Phase shifters are essential components in Butler matrices, implemented to provide the required phase difference between the antennas in the beam scanning arrays. For efficient beam control, the introduced phase shift must be stable over the operating bandwidth with low insertion loss, where several techniques have been investigated in the literature to achieve that property [47, 61]. One technique is based on using four port directional couplers with the isolated and through ports connected to each other. This produces the well-known Schiffman phase shifter for which the differential phase shift can be adjusted through careful selection of impedances of the coupled lines [62].

Figure 9a illustrates a Schiffman phase shifter based on PRGW technology, designed to achieve a 45^0 differential phase shift around 30 GHz [62]. Widths of the input and output lines, as well as those of the coupled section were all designed using

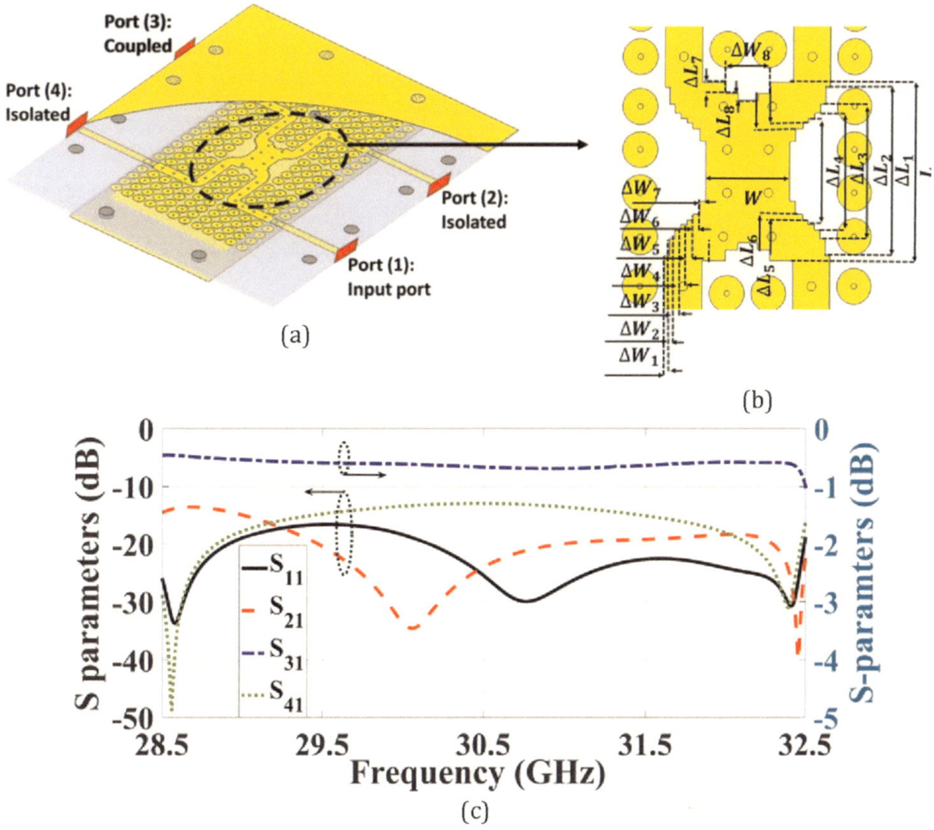

Figure 8.
Design of 0 dB coupler using rectangular coupling section to provide a crossover around 30 GHz. (a) 3D view of the crossover. (b) details of the coupling section. (c) simulated S-parameters assessing the isolation and coupling around 30 GHz.

Coupling Section		Step	1	2	3	4	5	6	7	8
W	8	ΔL_i	7.5	6	5.2	4.7	4.3	3.8	0.5	0.3
L	10	ΔW_i	0.22	0.24	0.26	0.31	0.25	0.33	0.26	0.92

Table 8.
Dimensions (in mm) of the coupling section and the matching steps in the crossover of ***Figure 8b****.*

Reference	Technology	Center frequency	Bandwidth	Insertion Loss (dB)	Size (λ^2)
[59]	SIW	20 GHz	10.5%	—	2
[60]	Slot SIW	30 GHz	16.7%	0.9	3.1
	Proposed PRGW	30 GHz	13.3%	0.5	2.25

Table 9.
Comparison of a proposed PRGW crossover with other technologies.

even\odd mode analysis of directional couplers [58]. The length of the coupled section was optimized for increased operating bandwidth. Final dimensions are given in **Table 10**, while the simulated S-parameters response is given in **Figure 9b**. The

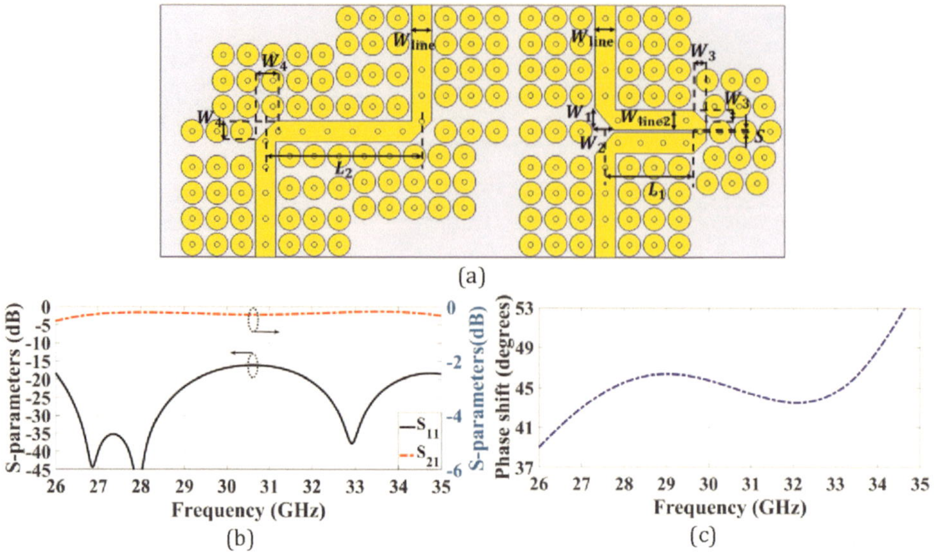

Figure 9.
Design and performance of the Schiffman 45° phase shifter. (a) detailed design dimensions. (b) simulated S-parameters magnitude. (c) phase difference between the input and output ports.

Parameter	W_{line}	W_{line2}	W_1	W_2	W_3	W_4	L_1	L_2	S
Value (mm)	1.38	1.35	1.49	1.28	0.88	1.4	5.99	10.64	0.15

Table 10.
Dimensions of the Schiffman phase shifter illustrated in ***Figure 9a****.*

device achieves 21.7% relative bandwidth at 30 GHz with 45^0phase shift $\pm 2.5^0$ differential phase error and less than 0.4 dB insertion loss [62].

3.4 Differential feeding power dividers

One of the main targets of mm-Wave beam switching arrays is to overcome the multipath fading in wireless communication channels by using space or polarization diversity techniques [9, 10]. These diversity techniques require the array to have highly stable radiation characteristic and low cross polarization level [61]. Such demands impose the usage of differential feeding for the array elements, which can be provided through out of phase power dividers. In this section, two designs for differential feeding power dividers are introduced.

The first design is shown in **Figure 10a**, where the power divider is implemented using two layers of PRGW structure coupled by I-shaped slot [63]. The first layer has the input feeding line connected to a matching stub optimized to achieve a deep matching level over a wide bandwidth, while the upper layer contains the two output PRGW lines, where the coupling through the I-shaped slot achieves the 180^0 phase difference [63]. The design dimensions are listed in **Table 11**, and the simulated S-parameters response shown in **Figure 10b** reveals a 20% relative bandwidth with more than 15 dB return loss and less than 0.3 dB insertion loss over the whole operating bandwidth. The phase difference between the output ports is stable around 180^0 ensuring differential output.

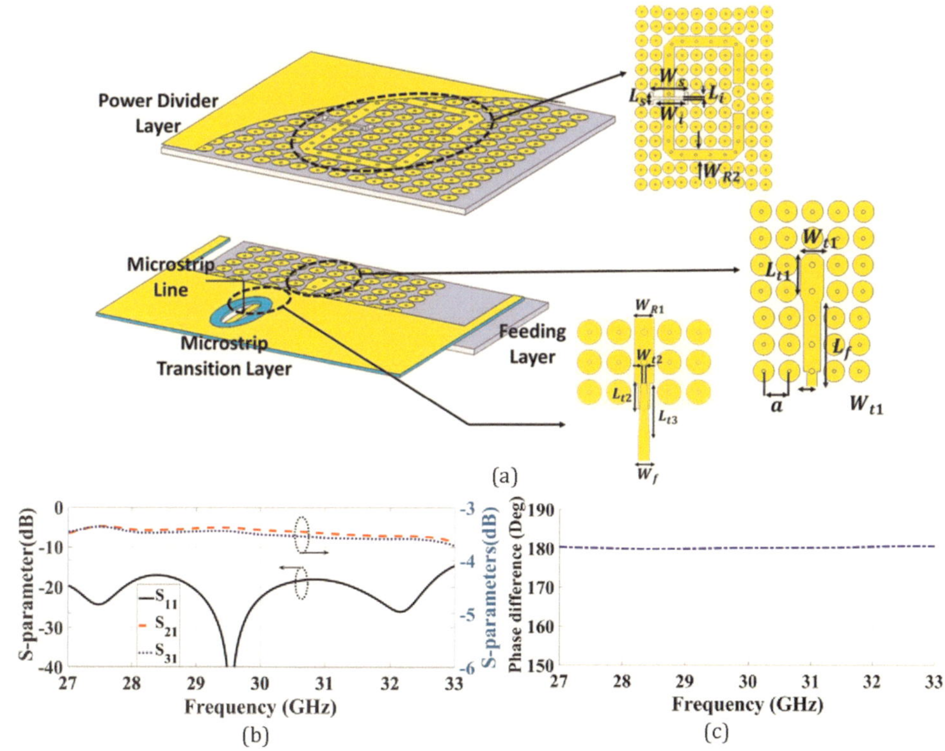

Figure 10.
Power divider design (a) detailed design dimensions for the feeding and divider layers. (b) Simulated S-parameters. (c) Phase difference between output ports.

Parameter	a	L_{t1}	W_{t1}	L_f	W_f	L_s	W_s
Value (mm)	1.8	2.85	1.6	5.5	0.65	1.9	5
Parameter	L_i	W_i	W_{R1}	W_{R2}	h_1	h_2	h_3
Value (mm)	0.34	4	1.37	1.37	0.5	0.5	0.127

Table 11.
Dimensions of the power divider in ***Figure 10a****.*

An alternative design is shown in **Figure 11a** where a hybrid ring or rat-race directional coupler is used to satisfy the power divider function with 180^0 in the output phase shift [64]. One advantage of rat-race couplers is that they can be used to produce in-phase or out-phase feeding according to the choice of the input port. The illustrated design has an introduced open circuited stub at the middle of the $3\lambda/4$ part of the ring which controls the signal splitting ratio between the output ports. Furthermore, at each port of the coupler, a quarter wavelength transformer is added to enhance the relative bandwidth [64]. The design dimensions of the ring, the stub and the quarter transformers are given in **Table 12**. With these dimensions, the device achieves 27.69% bandwidth at 30 GHz with more than 15 dB isolation as shown by the S-parameters in **Figure 11b**. A brief comparison is listed in **Table 13** among the former rat race coupler and other designs based on SIW technology, revealing the promising performance of the PRGW design.

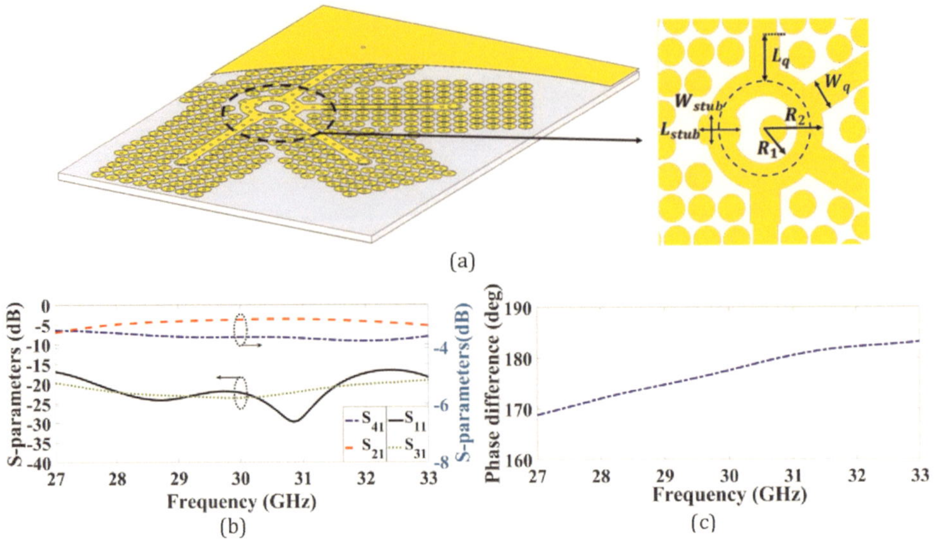

Figure 11.
Rat-race coupler design. (a) design with details and dimensions. (b) simulated S-parameters. (c) Phase difference between output ports.

Parameter	L_q	W_q	L_{stub}	W_{stub}	R_1	R_2
Value (mm)	2.5	1.7	2.4	0.8	1.85	3.35

Table 12.
Values of the rat-race coupler dimensions illustrated by ***Figure 11a****.*

Reference	Technology	Center frequency	Bandwidth	Return loss (dB)	Insertion loss (dB)	Amplitude balance (dB)
[65]	SIW	13 GHz	30%	18	20	3.35 ± 1.35
[66]	Half mode SIW	10.15 GHz	24.6%	12	15	3.8 ± 0.5
[67]	T-type folded SIW(TFSIW)	25.7 GHz	12.7%	20	20	4.3 ± 0.6
[68]	RSIW	8 GHz	12.5%	12	20	3.79 ± 0.5
	Proposed PRGW	30 GHz	27.9%	15	16.5	3.39 ± 0.5

Table 13.
Comparison with other technologies design for power dividers.

4. Antenna structures

The use of antenna arrays in communication handsets is feasible in mm-Wave communications due to the inherently small size of the antennas. However, applying diversity techniques in these systems require specific properties of the antennas, like stable and controllable radiation patterns, and the ability to produce a desired polarization with low cross polarization level. In this section, we present different antenna designs covering the possible polarizations, namely linear, circular, and dual polarizations as candidates for mm-Wave applications. The presented designs show promising performance in terms of beam stability, wide bandwidth, and low cross polarization level.

4.1 Linearly polarized antenna

As shown in **Figure 12**, a linearly polarized planar aperture antenna is presented, based on a similar design [69]. The antenna aperture has a cross-shaped patch in the middle, all in one top layer, fed by capacitive coupling from a differential feeding line in a bottom layer [58, 69]. The benefit of using planar radiating aperture is the ability to produce highly directive beam without need for increased dimensions. The feeding line is tapered and loaded with stubs to achieve acceptable matching level. The dimensions of the radiating element and the feeding structure are tuned to achieve optimum bandwidth, in terms of return loss and beam stability over the band [58]. The final dimensions are given in **Table 14**, while simulated and measured reflection coefficient and realized gain are plotted in **Figure 12c**. To obtain the measured results, the fabricated antenna was fed by the rat-race directional coupler, described in the previous section, to provide the differential feeding signal [64]. The results, as shown by **Figure 12c**, reveal a wide bandwidth

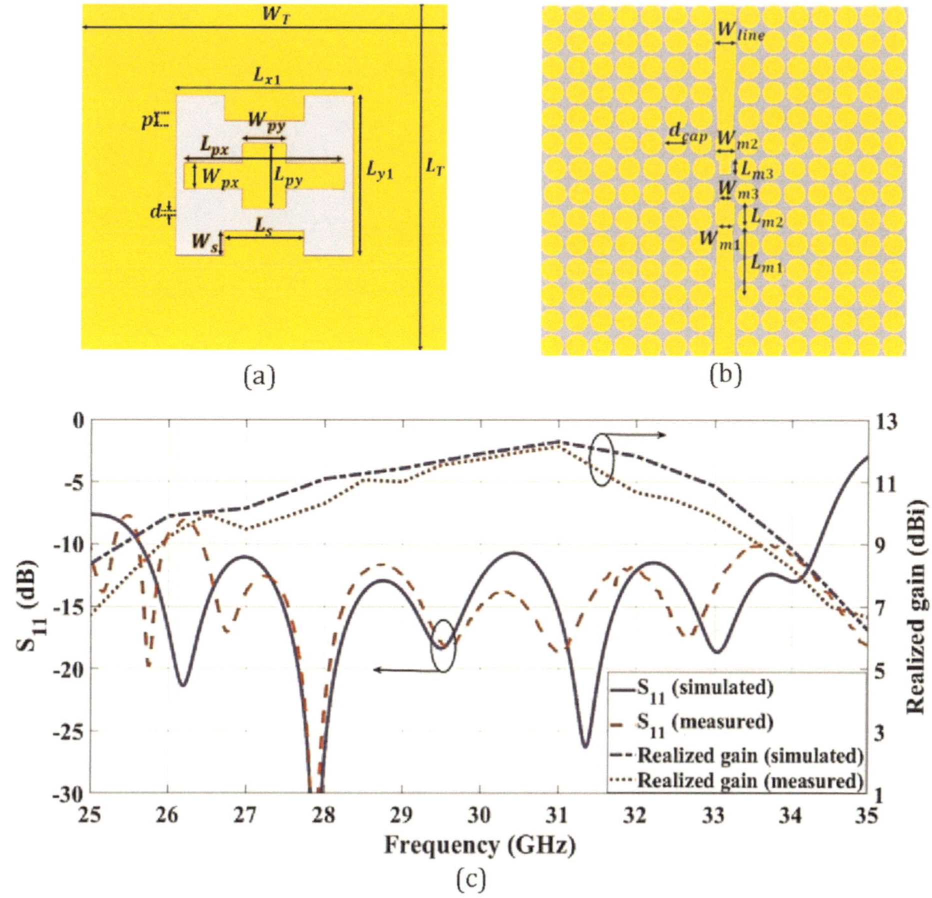

Figure 12.
Linearly polarized planar aperture antenna. (a) Top view. (b) bottom view. (c) simulated and measured S-parameter and realized gain.

Dimension	L_T	W_T	L_{x1}	L_{y1}	L_{px}	L_{py}	W_{px}	W_{py}	p	W_{line}
Value (mm)	23.8	25.5	12.45	10.94	11.09	4.36	1.7	2.97	0.84	1.38
Dimension	d	L_s	W_s	L_{m1}	L_{m2}	L_{m3}	W_{m1}	W_{m2}	W_{m3}	d_{cap}
Value (mm)	0.3	5.47	1.68	4.83	1.49	0.94	0.95	1.33	1.38	1.5

Table 14.
Dimensions of the planar aperture antenna and its differential feeding layer shown in ***Figure 12a*** *and* ***b****.*

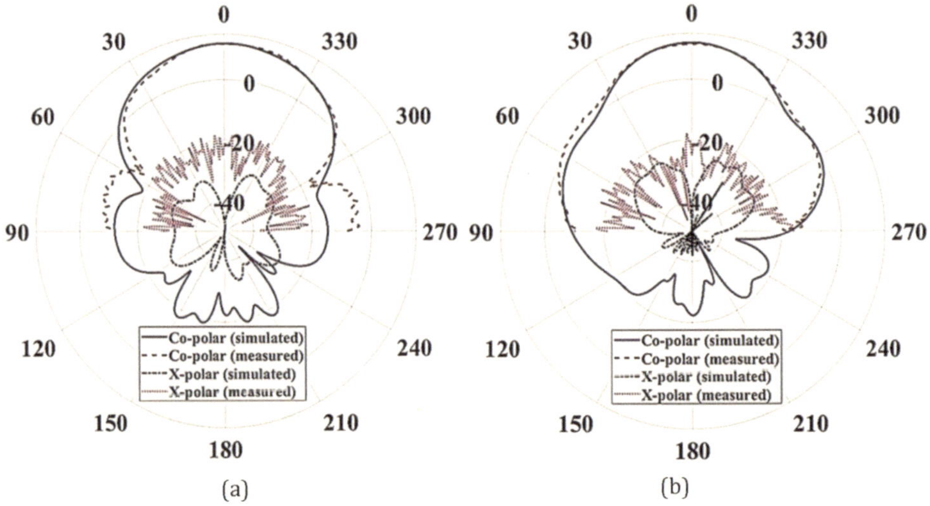

Figure 13.
Radiation properties of the linearly polarized planar aperture antenna. (a) simulated and measured co- and x- polarization patterns in E-plane at 30 GHz. (b) simulated and measured co- and x- polarization patterns in H-plane at 30 GHz.

over 25.6–34.3 GHz band with more than 10 dB return loss and with 12.28 dB$_i$ maximum gain. Moreover, the 3 dB gain bandwidth covers the range from 25.6 GHz up to 33.7 GHz, indicating a stable beam of the antenna. Beam stability is further revealed by measuring the radiation pattern at multiple frequencies over the band [58]. The simulated and measured patterns at the center frequency of 30 GHz are shown in **Figure 13a** and **b**. These patterns illustrate a very low cross polarization level in both E- and H- planes.

4.2 Circularly polarized antenna

Alternatively, the antenna design illustrated in **Figure 14** provides circular polarization (CP) at the same frequency band around 30 GHz. Unlike typical CP antenna designs, which depend on feeding the radiating element by two equal amplitude and quadrature phase signals, this design uses differential feeding to a planar aperture loaded with a polarizer [63]. The polarizer consists of an annular ring with two opposite cuts adjacent to the feeding position. These cuts perturb the excitation and cause the formulation of two orthogonal modes. The circular polarization is then adjusted through the circular patch added at the center to tune the amplitude and the phase relation of these orthogonal modes [63]. This tunning is performed by adjusting the patch size, and the orientation of the two introduced non-radiating edge slots. The antenna is designed on

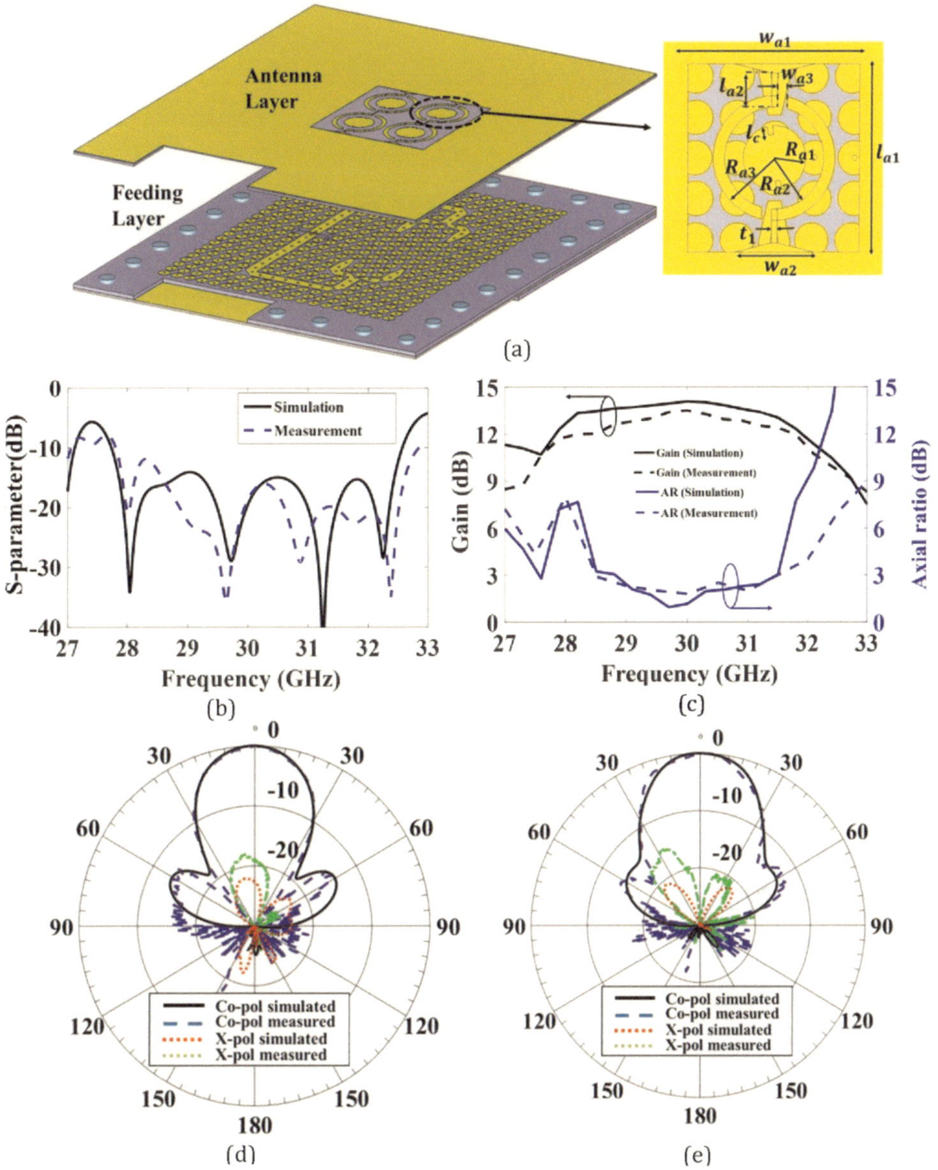

Figure 14.
Design and performance of the CP antenna: (a) design details. (b) simulated and measured $|S_{11dB}$. *(c) simulated and measured gain and axial ratio over the band. (d) simulated co- and x- polarization patterns (RH and LH CP's) in E-plane. (e) simulated co- and x- polarization patterns (RH and LH CP's) in H-plane.*

the Rogers RT5880 substrate with 0.127 mm thickness and 2.2 relative permittivity [63]. All the design dimensions are given in **Table 15**, for the planar aperture, the polarizer, and the patch in the center. Since this antenna is mainly aimed for use in communication arrays, the performance is investigated for an array of 4 elements shown by the 3D view of the design layers in **Figure 14a**. The simulated and measured reflection coefficients of that array are given in **Figure 14b** and indicate below −10 dB reflection over the 28–32.5 GHz frequency range, which is equivalent to 15% relative bandwidth at 30 GHz. Simulated and measured axial ratio illustrated in **Figure 14c** is showing 3 dB axial ratio

Dimension	w_{a1}	l_{a1}	w_{a2}	l_{a2}	w_{a3}	t_1	l_c	R_{a1}	R_{a2}	R_{a3}
Value (mm)	8.2	8.6	3.7	1.47	0.5	0.26	0.4	1.45	2.36	2.85

Table 15.
*Dimensions of the circularly polarized antenna illustrated by **Figure 14a**.*

over 28.5–31.5 GHz range, a slightly smaller relative bandwidth of 10%. However, these achievements are greater than the usually narrow bandwidths for CP antennas reported at 30 GHz. Regarding the cross polarization level and the beam angle, simulated and measured radiation patterns at 30 GHz are plotted in **Figure 14d** and **e**, showing less than −20 dB cross polarization levels at the direction of maximum radiation [63].

4.3 Dual polarized antenna

Finally, a dual polarized antenna is presented in **Figure 15**, where horizontal and vertical polarizations can be excited from two different isolated ports [70]. The two isolated ports are in different layers as shown in **Figure 15**. The antenna structure is implemented on the top layer and consists of two orthogonal magneto electric (ME) dipoles in the vertical and in the horizontal directions. Therefore, vertical and horizontal linear polarizations are obtained upon exciting the antenna. Each ME dipole is formed by means of two square patches, connected to the ground by five conducting plates at the patch corner [70]. This configuration, when excited properly, formulates an electric dipole from the plates, and a dual magnetic dipole from the vias. The structure in **Figure 15** has then two vertical and horizontal dipoles of each type [70].

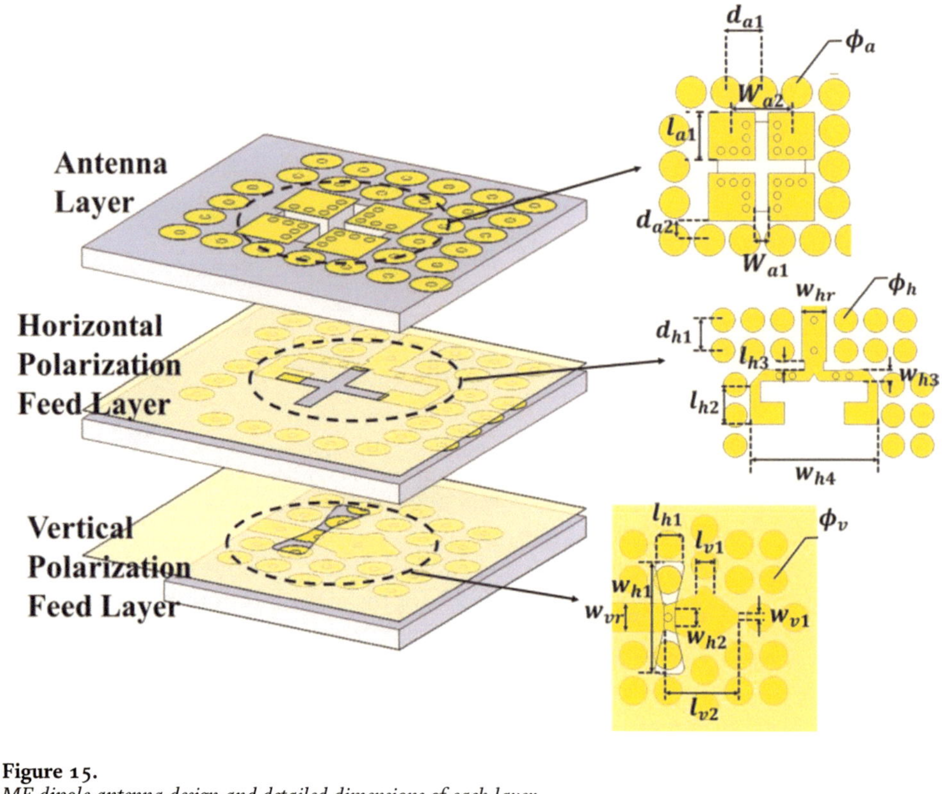

Figure 15.
ME dipole antenna design and detailed dimensions of each layer.

Parameter	w_{a1}	w_{a2}	w_{h1}	w_{h2}	w_{h3}	w_{h4}	w_{v1}	w_{vr}	w_{hr}	ϕ_a	ϕ_h
Value (mm)	0.5	4.1	5.6	1	0.75	7.3	0.4	1.37	1.37	1.6	1.3
Parameter	l_{h1}	l_{h2}	l_{h3}	l_{a1}	l_{v1}	l_{v2}	d_{a1}	d_{a2}	d_{h1}	ϕ_v	
Value (mm)	0.6	1.6	1.5	2.5	0.9	3.2	1.8	1	1.7	1.3	

Table 16.
Dimensions of the magneto-electric dipole antenna and its feeding structure illustrated in ***Figure 15****.*

Among the benefits of using ME dipole antennas are the directive pattern obtained identically in the E- and H-planes, and the stable beam of the antenna as to be illustrated in the subsequent results. **Table 16** summarizes the design dimensions of the ME dipole antenna and its feeding structure.

Figure 16a illustrates the simulated and measured S-parameters of this dual polarization ME dipole antenna. With respect to the self-matching of each port, the antenna achieves more than 10 dB return loss over 26.5–33.5 frequency range, which is approximately equivalent to 23% relative bandwidth at 30 GHz [70]. The isolation between the two ports exceeds 20 dB over the band. The simulated and measured gain values for each port are given in **Figure 16b**, indicating around 10 dB_i gain for each polarization [70]. The stability of the antenna beam is tested by measuring and simulating the radiation pattern at different frequency values, of which the results at 30 GHz are shown in **Figure 16c** and **d**. These drawn patterns are in E- and H- planes for the two co-polarizations along with their corresponding x-polarization level. The antenna achieves less than −20 dB x-polarization in each case, proving a great performance for polarization diversity applications [70].

5. Beamforming techniques

Beamforming is performed by means of a beam switching network (BSN) or a butler matrix which controls the signal feeding to\from the antennas in the array. The essential components constituting BSNs are the phase shifters, hybrid couplers, power dividers and crossovers discussed in the former sections. Upon introducing these components, we present two designs of butler matrices in this section and the deployment of these matrices to design beam scanning antenna arrays.

Figure 17a shows a 4 × 4 butler matrix realized by four directional couplers to distribute the signals, with the desired phase relations, between the four feeding ports and the four array elements [71]. Various designs of hybrid couplers are presented in Section 3.1, which cover the whole desired band of 27–33 GHz. This 4 × 4 butler matrix achieves two-dimensional beam scanning in the horizontal plane. The obtainable beam directions are $\theta_0 = 45^0$ and $\phi_0 = -45^0, 45^0, 135^0$ and -135^0 for excitations from Ports 1, 2, 3 and 4 respectively. The performance of this matrix is first evaluated in terms of ports' matching and isolation. Simulated and measured scattering parameters of Port 1 are shown in **Figure 17b** where more than 10 dB return loss and isolation is obtained over the band. These levels are then expected for all other ports due to the design symmetry. To test the scanning abilities, the matrix is used to feed an array of four ME-dipole antennas and the radiation pattern is simulated and measured [71]. The array shows a measured realized gain of 9.7 ±0.4 dB_i where the reduction from the simulated results is mainly caused by feeding network losses. The simulated and measured radiation patterns in $\phi = -45^0$ and $\phi = 135^0$ planes are given for Port 1 and Port 3

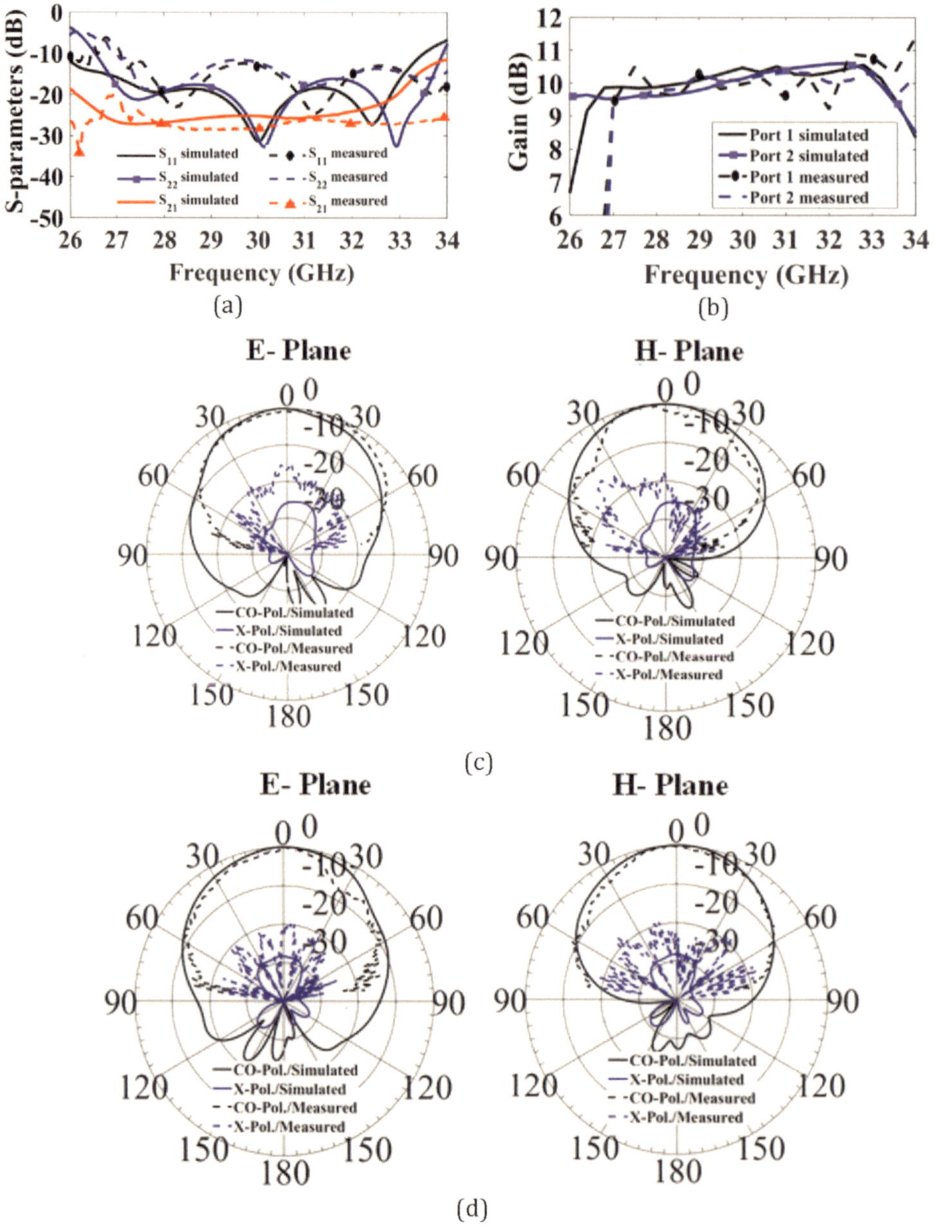

Figure 16.
Magneto-electric dipole antenna performance. (a) simulated and measured S-parameters. (b) simulated and measured gain over the band. (c) simulated and measured radiation patterns at 30 GHz for horizontal polarization in E- and H-planes. (d) simulated and measured radiation patterns at 30 GHz for vertical polarization in E- and H-planes respectively.

excitations. The beam is nearly identical, emphasizing efficient scanning, with low cross-polarization level of less than −20 dB at the main direction [71].

An alternative 4 × 4 butler matrix is shown in **Figure 18a**, where the ring power divider, cascaded crossover and the Schiffman phase shifter described in the previous sections are all integrated in one network [51]. The four input ports to the matrix are all coaxial ports where coaxial to PRGW transition are used. The matrix is feeding four

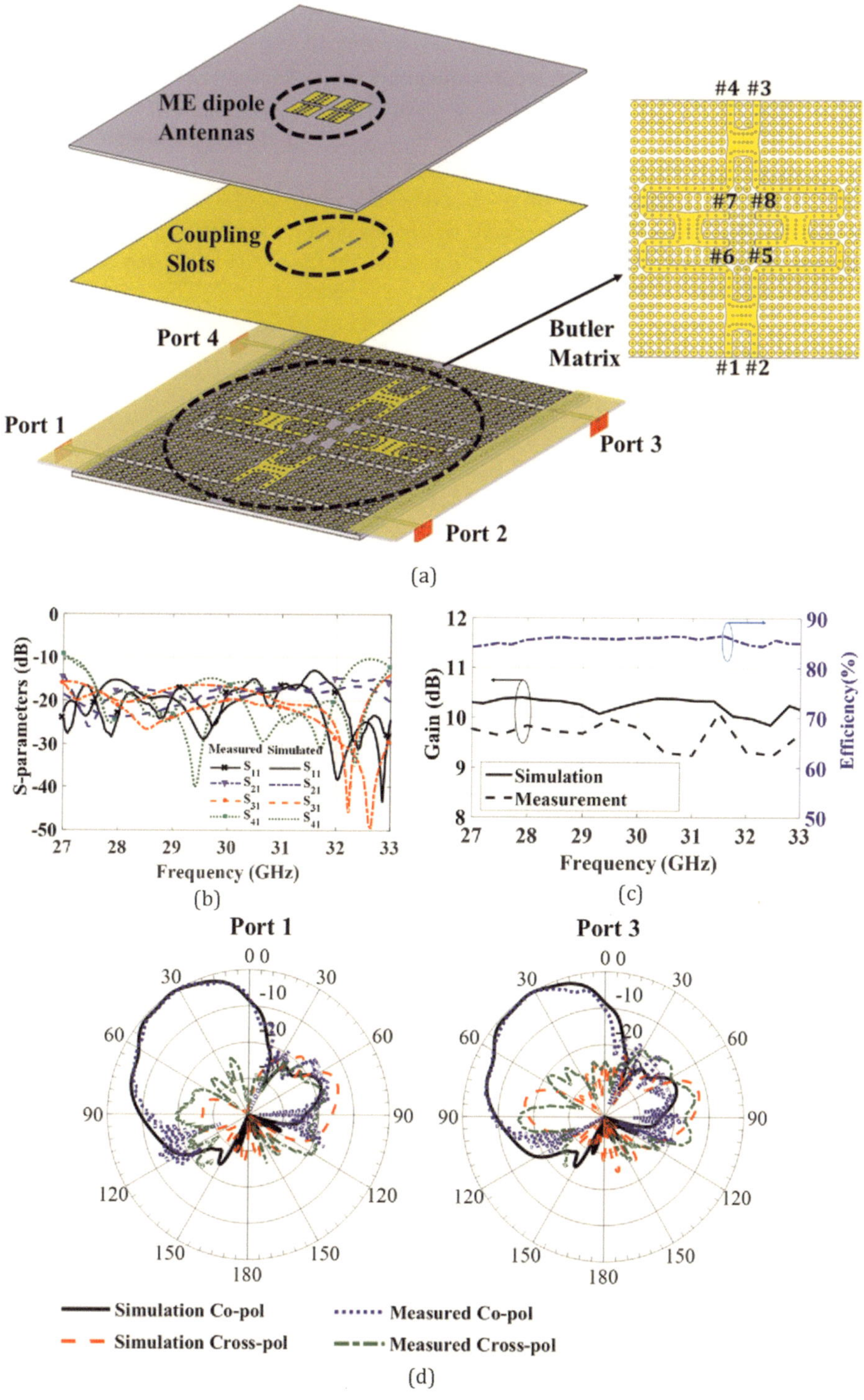

Figure 17.
4×4 butler matrix, realized by four directional couplers, and achieves 2D beam scanning. (a) detailed 3D view of the matrix and the fed antennas. (b) simulated and measured reflection coefficient and isolation of the first port. (c) simulated and measured gain and efficiency over the band. (d) simulated and measured radiation pattern at $\phi = -45°$ with Port 1 excited (left) and at $\phi = 135°$ with Port 3 excited (right).

wideband semi-log periodic dipole antennas to test the array performance [72]. Since the radiating elements are all printed antennas with microstrip feeding input, MSPL to PRGW transition is used at each matrix output. Two secondary antennas are added to enhance the symmetry of the radiation pattern.

The array performance is illustrated by **Figure 18b–d**. The simulated and measured scattering parameters indicate isolation and return loss levels of more than 10 dB over the frequency range of 26–34 GHz. The array achieves radiation efficiency of 78% with gain value ranging from 10 dB_i to 11.35 dB_i over the whole frequency band (for excitation from port 1). The obtainable angles of the beam are −13, −36, 36, and 13° for excitation from ports 1, 2, 3, and 4 respectively [51]. The simulated and measured gain patterns in **Figure 18d** are obtained at 30 GHz and emphasize the scanning ability of the array.

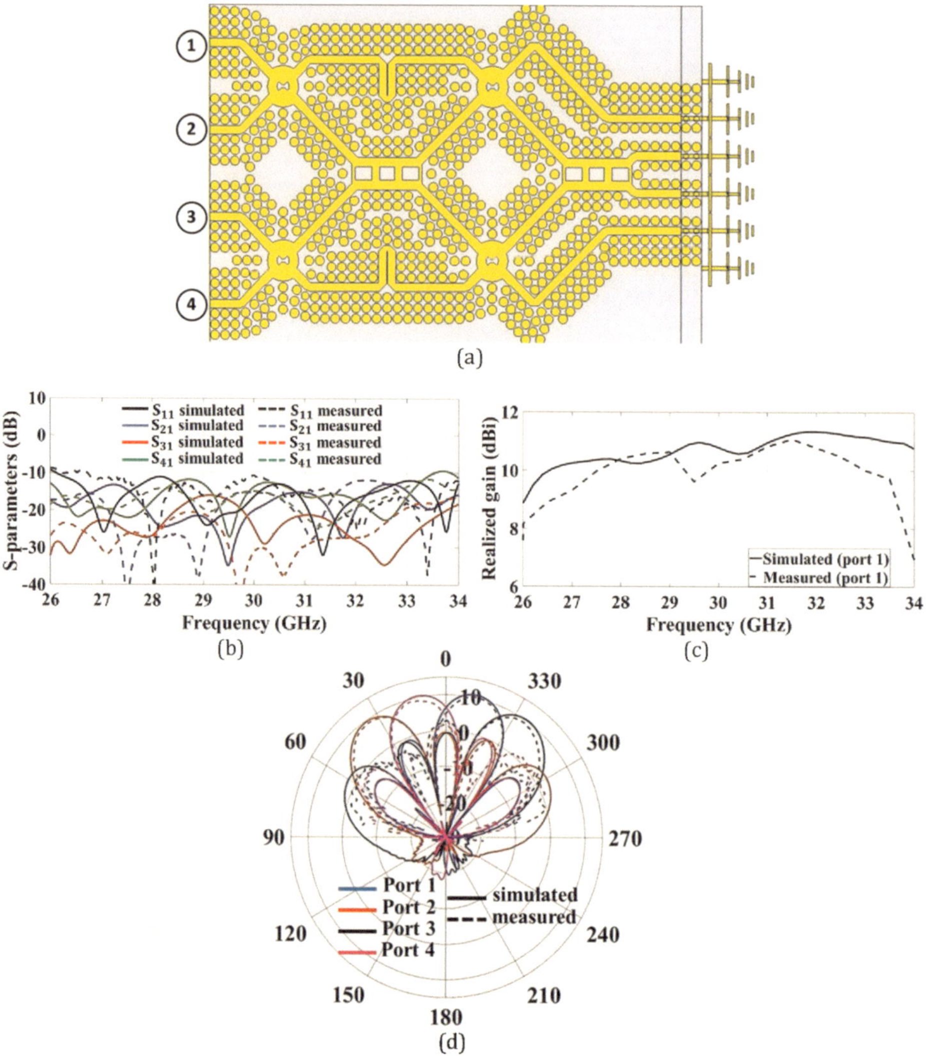

Figure 18.
Design and performance of the second 4 × 4 butler matrix for beam switching network. (a) design details. (b) simulated and measured reflection and isolation for Port 1. (c) simulated and measured realized gain for Port 1. (d) simulated and measured radiation pattern for the excitation from the four ports.

6. Conclusions

Printed ridge gap waveguide is an emerging technology with powerful capabilities that allow designing efficient beam switching networks for mm-Wave communications. In this chapter, various PRGW components designed for beam switching arrays at 30 GHz are presented. First, we present transitions to\from other guiding standards like coaxial and MSPL structures. These transitions provide efficient integration with these guiding standards. Then, hybrid couplers with rectangular and circular coupling sections are introduced. Crossover designs are also provided with various design techniques; either using multistage directional couplers or using compact single stage stepped design. A Schiffman phase shifter with 45^0 phase shift is also presented. Differential power dividers are introduced, one relying on slot coupling to two identical lines and the other depending on rat race design. The chapter also presents different designs of antennas covering all possible polarizations. It is then revealed how the usage of PRGW technology can provide efficient excitation of antennas with different polarizations and design profiles. Full beam switching networks are presented in the last section, where two different 4×4 butler matrices are introduced with their usage in achieving the beam scanning function of the array. Assessment of these various devices reveals the promising performance of the new PRGW structure over other competing technology candidates. The new PRGW technology can give better performance in terms of device size, bandwidth and fabrication costs.

Author details

Mohammad Ali AbdElraheem[1], Mohamed Mamdouh M. Ali[2]*, Islam Afifi[3] and Abdel R. Sebak[1]

1 Concordia University, Montreal, Canada

2 Assiut University, Asyut, Egypt

3 Cairo University, Cairo, Egypt

*Address all correspondence to: mohamed.ali@ieee.org

IntechOpen

© 2022 The Author(s). Licensee IntechOpen. This chapter is distributed under the terms of the Creative Commons Attribution License (http://creativecommons.org/licenses/by/3.0), which permits unrestricted use, distribution, and reproduction in any medium, provided the original work is properly cited.

References

[1] ITU-T Recommendations and other publications [Internet]. ITU. 2022. Available from: https://www.itu.int:443/en/ITU-T/publications/Pages/default.aspx

[2] IEEE Standard for Air Interface for Broadband Wireless Access Systems. IEEE Std 80216-2012 Revis IEEE Std 80216-2009. 2012;1–2542

[3] Tehrani RH, Vahid S, Triantafyllopoulou D, Lee H, Moessner K. Licensed spectrum sharing schemes for mobile operators: A survey and outlook. IEEE Communications Surveys and Tutorials. 2016;**18**(4):2591-2623

[4] Jameel F, Hamid Z, Jabeen F, Zeadally S, Javed MA. A survey of device-to-device communications: Research issues and challenges. IEEE Communications Surveys and Tutorials. 2018;**20**(3):2133-2168

[5] Chen R, Long WX, Mao G, Li C. Development trends of mobile communication systems for railways. IEEE Communications Surveys and Tutorials. 2018;**20**(4):3131-3141

[6] Baltaci A, Dinc E, Ozger M, Alabbasi A, Cavdar C, Schupke D. A survey of wireless networks for future aerial communications (FACOM). IEEE Communications Surveys and Tutorials. 2021;**23**(4):2833-2884

[7] Xia N, Chen HH, Yang CS. Radio resource management in machine-to-machine communications—A survey. IEEE Communications Surveys and Tutorials. 2018;**20**(1):791-828

[8] Agiwal M, Roy A, Saxena N. Next generation 5G wireless networks: A comprehensive survey. IEEE Communications Surveys and Tutorials. 2016;**18**(3):1617-1655

[9] Xiao M, Mumtaz S, Huang Y, Dai L, Li Y, Matthaiou M, et al. Millimeter wave communications for future mobile networks. IEEE Journal on Selected Areas in Communications. 2017 Sep;**35**(9):1909-1935

[10] Rappaport TS, Sun S, Mayzus R, Zhao H, Azar Y, Wang K, et al. Millimeter wave mobile communications for 5g cellular: It will work! IEEE Access. 2013;**1**:335-349

[11] Rappaport TS, Murdock JN, Gutierrez F. State of the Art in 60-GHz integrated circuits and systems for wireless communications. Proceedings of the IEEE. 2011 Aug;**99**(8):1390-1436

[12] Chettri L, Bera R. A comprehensive survey on internet of things (IoT) toward 5G wireless systems. IEEE Internet of Things Journal. 2020;**7**(1):16-32

[13] Meinel HH. Commercial applications of millimeterwaves: History, present status, and future trends. IEEE Transactions on Microwave Theory and Techniques. 1995;**43**(7):1639-1653

[14] Pereira de Figueiredo FA. An overview of massive MIMO for 5G and 6G. IEEE Latin America Transactions. 2022;**20**(6):931-940

[15] Wang X, Kong L, Kong F, Qiu F, Xia M, Arnon S, et al. Millimeter wave communication: A comprehensive survey. IEEE Communications Surveys and Tutorials. 2018;**20**(3):1616-1653

[16] Rangan S, Rappaport TS, Erkip E. Millimeter-wave cellular wireless networks: Potentials and challenges.

Proceedings of the IEEE. 2014;**102**(3): 366-385

[17] Shafi M, Molisch AF, Smith PJ, Haustein T, Zhu P, De Silva P, et al. 5G: A tutorial overview of standards, trials, challenges, deployment, and practice. IEEE Journal on Selected Areas in Communications. 2017;**35**(6):1201-1221

[18] Rappaport TS, Gutierrez F, Ben-Dor E, Murdock JN, Qiao Y, Tamir JI. Broadband millimeter-wave propagation measurements and models using adaptive-beam antennas for outdoor urban cellular communications. IEEE Transactions on Antennas and Propagation. 2013;**61**(4):1850-1859

[19] Gupta A, Jha RK. A survey of 5G network: Architecture and emerging technologies. IEEE Access. 2015;**3**: 1206-1232

[20] Tataria H, Shafi M, Molisch AF, Dohler M, Sjöland H, Tufvesson F. 6G wireless systems: Vision, requirements, challenges, insights, and opportunities. Proceedings of the IEEE. 2021;**109**(7): 1166-1199

[21] Di Renzo M, Haas H, Ghrayeb A, Sugiura S, Hanzo L. Spatial modulation for generalized mimo: Challenges, opportunities, and implementation. Proceedings of the IEEE. 2014;**102**(1): 56-103

[22] Islam SMR, Avazov N, Dobre OA, Kwak K, sup. Power-domain non-orthogonal multiple access (NOMA) in 5G systems: Potentials and challenges. IEEE Communications Surveys and Tutorials. 2017;**19**(2):721-742

[23] Natarajan A, Reynolds SK, Tsai MD, Nicolson ST, Zhan JHC, Kam DG, et al. A fully-integrated 16-element phased-array receiver in SiGe BiCMOS for 60-GHz communications. IEEE Journal of Solid-State Circuits. 2011;**46**(5): 1059-1075

[24] Garg R, Natarajan AS. A 28-GHz low-power phased-array receiver front-end with 360° RTPS phase shift Range. IEEE Transactions on Microwave Theory and Techniques. 2017;**65**(11):4703-4714

[25] Moulder WF, Khalil W, Volakis JL. 60-GHz two-dimensionally scanning array employing wideband planar switched beam network. IEEE Antennas and Wireless Propagation Letters. 2010; **9**:818-821

[26] Ashraf N, Kishk AA, Sebak A. Broadband millimeter-wave beamforming components augmented with AMC packaging. IEEE Microwave and Wireless Components Letters. 2018; **28**(10):879-881

[27] Denlinger EJ. Losses of microstrip lines. IEEE Transactions on Microwave Theory and Techniques. 1980;**28**(6): 513-522

[28] Wu K, Bozzi M, Fonseca NJG. Substrate integrated transmission lines: Review and applications. IEEE Journal of Microwaves. 2021;**1**(1):345-363

[29] Lai Q, Fumeaux C, Hong W, Rü V. Characterization of the propagation properties of the half-mode substrate integrated waveguide. IEEE Transactions on Microwave Theory and Techniques. 2009;**57**(8):1996-2004

[30] Cassivi Y, Perregrini L, Arcioni P, Bressan M, Wu K, Conciauro G. Dispersion characteristics of substrate integrated rectangular waveguide. IEEE Microwave and Wireless Components Letters. 2002;**12**(9):333-335

[31] Xu F, Wu K. Guided-wave and leakage characteristics of substrate integrated waveguide. IEEE

Transactions on Microwave Theory and Techniques. 2005;**53**(1):66-73

[32] Kildal PS, Alfonso E, Valero-Nogueira A, Rajo-Iglesias E. Local metamaterial-based waveguides in gaps between parallel metal plates. IEEE Antennas and Wireless Propagation Letters. 2009;**8**:84-87

[33] Bayat-Makou N, Kishk AA. Realistic air-filled tem printed parallel-plate waveguide based on ridge gap waveguide. IEEE Transactions on Microwave Theory and Techniques. 2018;**66**(5):2128-2140

[34] Sharifi Sorkherizi M, Kishk AA. Fully printed gap waveguide with facilitated design properties. IEEE Microwave and Wireless Components Letters. 2016;**26**(9):657-659

[35] Farahani M, Akbari M, Nedil M, Denidni TA, Sebak AR. A novel low-loss millimeter-Wave 3-dB 90° ridge-gap coupler using large aperture progressive phase compensation. IEEE Access. 2017; **5**:9610-9618

[36] Sifat SM, Ali MMM, Shams SI, Sebak AR. High gain bow-tie slot antenna array loaded with grooves based on printed ridge gap waveguide technology. IEEE Access. 2019;**7**:36177-36185

[37] Dadgarpour A, Bayat-Makou N, Antoniades MA, Kishk AA, Sebak A. A dual-polarized magnetoelectric dipole array based on printed ridge gap waveguide with dual-polarized split-ring resonator lens. IEEE Transactions on Antennas and Propagation. 2020;**68**(5): 3578-3585

[38] Zhao Z, Denidni TA. Millimeter-wave printed-RGW hybrid coupler with symmetrical square feed. IEEE Microwave and Wireless Components Letters. 2020;**30**(2):156-159

[39] Kakhki MB, Dadgarpour A, Antoniades MA, Sebak AR, Denidni TA. Magneto-electric dipole antennas loaded with meta-lens for 5G MIMO pattern diversity applications. IEEE Transactions on Antennas and Propagation. 2021;**1–1**

[40] Ali MMM, Al-Hasan M, Mabrouk IB, Denidni TA. Ultra-wideband hybrid magneto-electric dielectric-resonator dipole antenna fed by a printed RGW for millimeter-wave applications. IEEE Access. 2022;**10**:2028-2036

[41] krn5. CST Studio Suite 3D EM Simulation and Analysis Software [Internet]. 2022. Available from: https://www.3ds.com/products-services/simulia/products/cst-studio-suite/

[42] Sharifi Sorkherizi M, Kishk Ahmed A. Transition from Microstrip to Printed Ridge Gap Waveguide for Millimeter-Wave Application. Vancouver, BC, Canada: IEEE; 2015. pp. 1588-1589

[43] Pourahmadazar J, Farahani M, Denidni T. Printed Ridge Gap Waveguide Rotman Lens for Millimetre-wave Applications. 2018 18th International Symposium on Antenna Technology and Applied Electromagnetics (ANTEM); 2018. pp. 1-2

[44] Ali MMM. Millimeter-Wave Components and Antennas for Spatial and Polarization Diversity using PRGW Technology [PhD]. Montr'eal, Qu'ebec, Canada: Concordia University; 2020

[45] Ali MMM, Shams SI, Sebak AR. Printed ridge gap waveguide 3-dB coupler: Analysis and design procedure. IEEE Access. 2018;**6**:8501-8509

[46] Afifi I, Ali MMM, Sebak AR. Analysis and design of a wideband coaxial transition to metal and printed ridge gap waveguide. IEEE Access. 2018; **6**:70698-70706

[47] Han S, Lin IC, Xu Z, Rowell C. Large-scale antenna systems with hybrid analog and digital beamforming for millimeter wave 5G. IEEE Communications Magazine. 2015;**53**(1): 186-194

[48] Ali MMM, Shams SI, Elsaadany M, Sebak A. Printed RGW 3-dB Backward Coupler for Millimeter Wave Applications. In Montreal, QC, Canada: IEEE; 2020

[49] Ali MMM, Shams SI, Sebak A. Ultra-wideband printed ridge gap waveguide hybrid directional coupler for millimetre wave applications. IET Microwaves, Antennas & Propagation. 2019;**13**(8):1181-1187

[50] MMM A, Denidni TA. A Wideband Millimeter-Wave 3-dB Hybrid Coupler Based on Printed-RGW Technology. Singapore, Singapore: IEEE; 2021

[51] Afifi I, Sebak AR. Wideband 4 × 4 butler matrix in the printed ridge gap waveguide technology for millimeter-wave applications. IEEE Transactions on Antennas and Propagation. 2020;**68**(11): 7670-7675

[52] Hildebrand LT. Results for a simple compact narrow-wall directional coupler. IEEE Microwave and Guided Wave Letters. 2000;**10**(6):231-232

[53] Djerafi T, Wu K. Super-compact substrate integrated waveguide cruciform directional coupler. IEEE Microwave and Wireless Components Letters. 2007;**17**(11):757-759

[54] Liu B, Hong W, Zhang Y, Tang HJ, Yin X, Wu K. Half mode substrate integrated waveguide 180° 3-db directional couplers. IEEE Transactions on Microwave Theory and Techniques. 2007;**55**(12):2586-2592

[55] Afifi I, Ali MMM, Sebak AR. Analysis and Design of a 30 GHz Printed Ridge Gap Ring-Crossover. Atlanta, GA, USA: IEEE; 2019. pp. 65-66

[56] Ali MMM, Afifi I, Sebak AR. Design of Printed RGW Crossover for Millimeter Wave Beam Switching Network. Atlanta, GA, USA: IEEE; 2019. pp. 63-64

[57] Ali MMM, Sebak A. Compact printed ridge gap waveguide crossover for future 5G wireless communication system. IEEE Microwave and Wireless Components Letters. 2018;**28**(7):549-551

[58] Afifi I. 30 GHz Printed Ridge Gap Components and Antennas for Imaging Systems [PhD]. Montreal, QC, Canada: Concordia University; 2020

[59] Murai K, Ikeuchi H, Kawai T, Kishihara M, Ohta I. Broadband design method of SIW directional couplers. 2011 China-Japan Joint Microwave Conference; 2011. pp. 1-4

[60] Zheng SY, Ye XF. Ultra-Compact Wideband Millimeter-Wave Crossover Using Slotted SIW Structure. Nanjing, China: IEEE; 2016. pp. 1-2

[61] Kutty S, Sen D. Beamforming for millimeter wave communications: An inclusive survey. IEEE Communications Surveys and Tutorials. 2016;**18**(2): 949-973

[62] Afifi I, Sebak AR. Wideband Printed Ridge Gap 45° Schiffman Phase Shifter for Millimeter Wave Systems. Montreal, QC, Canada: IEEE; 2020. pp. 101-102

[63] Ali MMM, Sebak A. Printed RGW circularly polarized differential feeding antenna array for 5G communications. IEEE Transactions on Antennas and Propagation. 2019 May;**67**(5): 3151-3160

[64] Afifi I, Sebak AR. Wideband printed ridge gap rat-race coupler for differential feeding antenna. IEEE Access. 2020;**8**: 78228-78235

[65] Dehdasht-Heydari R, Forooraghi K, Naser-Moghadasi M. Efficient and accurate analysis of a substrate integrated waveguide (SIW) rat-race coupler excited by four U-shape slot-coupled transitions. Applied Computational Electromagnetics Society Journal. 2015 Jan;**1**(30):42-49

[66] Zou X, Tong CM, Li CZ, Pang WJ. Wideband hybrid ring coupler based on half-mode substrate integrated waveguide. IEEE Microwave and Wireless Components Letters. 2014 Sep; **24**(9):596-598

[67] Ding Y, Wu K. Miniaturized Hybrid Ring Circuits Using T-type Folded Substrate Integrated Waveguide (TFSIW). Boston, MA, USA: IEEE; 2009. pp. 705-708

[68] Ali AAM, El-Shaarawy HB, Aubert H. Miniaturized hybrid ring coupler using electromagnetic bandgap loaded ridge substrate integrated waveguide. IEEE Microwave and Wireless Components Letters. 2011 Sep; **21**(9):471-473

[69] Afifi I, Alzidani M, Sebak A. Wideband Printed Ridge Gap Waveguide Differential Feeding Aperture Antenna for Millimeter Wave Applications. Atlanta, GA, USA: IEEE; 2019. pp. 267-268

[70] Ali MMM, Afifi I, Sebak AR. A dual-polarized magneto-electric dipole antenna based on printed ridge gap waveguide technology. IEEE Transactions on Antennas and Propagation. 2020 Nov;**68**(11): 7589-7594

[71] Ali MMM, Sebak AR. 2-D scanning magnetoelectric dipole antenna array fed by RGW butler matrix. IEEE Transactions on Antennas and Propagation. 2018 Nov;**66**(11):6313-6321

[72] Afifi I, Ali MMM, Sebak AR. Wideband Printed Ridge Gap Semi-Log Periodic Structure Antenna for Millimeter Wave Applications. 2018 18th International Symposium on Antenna Technology and Applied Electromagnetics (ANTEM); 2018. pp. 1-2

Section 2

Additive Manufacturing

Chapter 5

Manufacturing Methods Based on Planar Circuits

Darío Herraiz, Leticia Martínez, José A. Ballesteros, Marcos D. Fernandez, Héctor Esteban and Ángel Belenguer

Abstract

Manufacture of hybrid 3D-planar circuits, especially those incorporating empty waveguides on substrates, can benefit from most standardized planar fabrication processes, although they are not exactly the same. For this reason, planar circuit manufacturing methods must be adapted to the requirements of these new circuits. Through numerous fabrications and successful designs, several enhancing strategies have been established to improve all the manufacturing phases to achieve better results. They all have been proved in the following substrate-integrated technologies for the manufacturing of microwave devices: ESIW, ESICL, continuous profile, and microstrip. Thanks to these improvements, good-quality prototypes such as transitions, filters, circulators, couplers, antennas, among others, have been fabricated. Throughout the next chapter, these strategies applied along the manufacturing process will be explained: from the first manufacturing phase to the final welding of the whole circuit and taking into account external elements such as wires that may be present in these structures. For this purpose, some devices that have been published will be used as examples.

Keywords: ESIW, ESICL, waveguide, microstrip, planar, 3D, manufacturing, fabrication, substrate integrated circuits, coaxial

1. Introduction

Planar circuit manufacturing processes have a long-term way. Many of the fabrication techniques of planar circuits can be applied to build 3D structures (substrate-integrated waveguides—SIW [1], empty substrate-integrated waveguides—ESIW [2], ridge empty substrate-integrated waveguides—RESIW [3], double-ridge empty substrate-integrated waveguides—DRESIW [4], substrate-integrated coaxial line—SICL [5] and empty substrate-integrated coaxial line—ESICL [6] or a combination with others as a continuous profile [7]) piling up layers of planar substrates.

In this chapter, some strategies and techniques adapted from planar to planar-3D structures will be explained with examples of manufactured prototypes, this philosophy being ideal for microwave prototypes (300 MHz–300 GHz), because the sizes of cavities are small enough to build heaping up with few small height substrates, of no more than 1.5 mm.

The prototypes can be fed with planar circuits such as coplanar or microstrip lines; because of this, at some point the feeding planar line needs to be connected with the inner waveguide or the connector through transitions, which needs to be also taken into account. One of the most typical transitions to do this are tapers, that is, transitions with a progressive geometrical variation of the guide which replicates the 3D launchers but planarly [8, 9]. Another way to do this is through wires [10].

2. Machines

The whole manufacturing procedure needs machines to fabricate each layer and measuring systems to check each manufacturing step.

2.1 Manufacturing machines

To manufacture the different layers, some machines must be used. A few possible ones are explained in the following paragraphs.

2.1.1 Plotter/cutter laser

This kind of machines use lasers to mill, cut, and drill substrates. Lasers are usually conical, so that the incidence angle is not 90°, but slightly lower, being necessary to minimize this effect as it will be explained in section 4.3. Laser machines are slower than mechanical milling machines, but they have a higher precision. In **Figure 1**, an example of these machines can be seen.

2.1.2 Milling machine

These machines use different mechanical drilling tools to mill, cut, and drill. The main downsides are the precision in comparison with laser ones and corner inner cuts, which are rounded because of the cylindrical shape of this tools. Although milling machines are very useful, they have some limitations: First, milling drills sizes vary from 0.2 mm to 3 mm, and their intermediates' values depend on the commercial drilling tools. In **Figure 2**, an example of these machines can be seen.

Both laser and milling machines can work independently or combined, given that mechanical machines are much faster than laser ones, especially in the drilling processes; being good enough for those kind of prototypes or parts in which drilling accuracy is not crucial. The combination of both machines decreases hugely the total production time without decreasing the accuracy, and it is highly recommended for prototypes with rigorous requirements. Resolutions of high-performance machines shift from 0.1 μm to 2 μm, and their repeatability oscillates from $\pm$ 0.1 μm to $\pm$ 2 μm; it depends on the machine. Theoretically, both machines can reach similar resolutions; however, drilling tools tend to be slightly bigger/smaller reducing resolution till 100 $\pm$ μm.

2.1.3 Multipress

These machines press the layers of prototypes and heat them up to high temperatures, melting the soldering paste to weld the prototype. Temperature, pressure, and

Figure 1.
Plotter/cutter laser machine.

time can be configured among a wide spectrum of values. In **Figure 3**, an example of these machines can be seen.

2.1.4 Reflow oven

Reflow ovens heat the prototype up to high temperatures to melt the welding paste and solder the prototype. The pressure must be applied with auxiliar elements such as

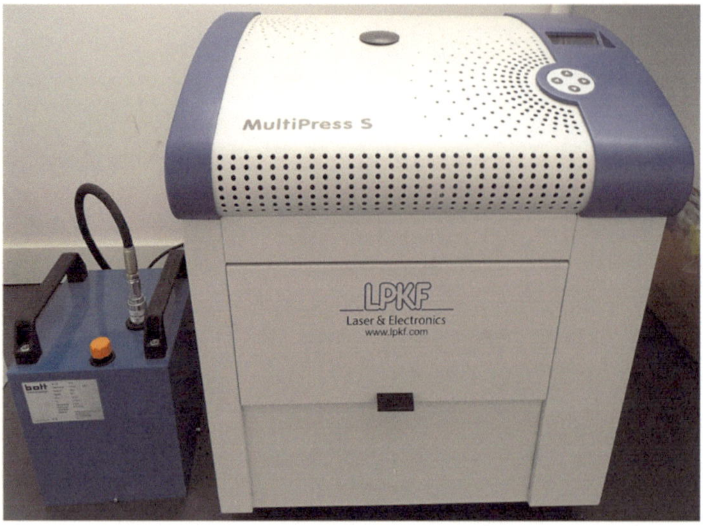

Figure 2.
Multipress machine.

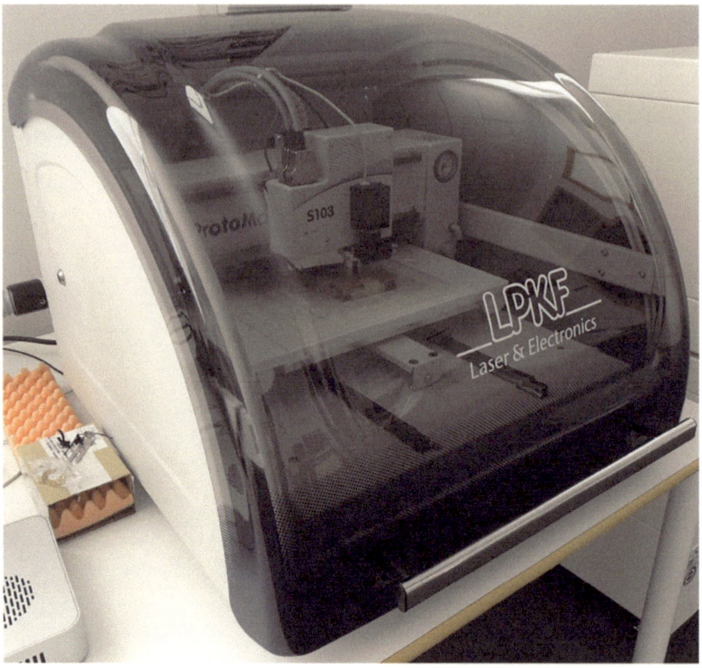

Figure 3.
Milling machine.

screws and nuts. Temperature can be selected to fit the recommended values and times for each soldering paste, this oven reaches around 270 °C (it can be higher depending on the specific model) with 1°C resolution steps. In **Figure 4**, an example of these machines can be seen.

Figure 4.
Reflow oven.

2.1.5 Soldering plate

These hot plate systems have a flat area to heat prototypes and allow manipulation at the same time. They are very useful combined with welders to manually solder specific parts of the structure. In **Figure 5**, an example of these machines can be seen.

2.1.6 Welder

These tools heat up fastly small areas by contacting with the welding head, allowing soldering paste and welding wires to melt. In **Figure 6**, an example of these machines can be seen.

2.1.7 Galvanic metallizer for PCBs

This machine adds a layer of copper to the PCB, being necessary to metallize holes and cuts in the prototype. The layer of metal can be estimated mathematically depending on temperature, time, and current, or it could be measured if necessary. Maximum dimensions for substrates are around 200×290 mm/230 mm $\times$ 330 mm for the typical models. In **Figure 7**, an example of these machines can be seen.

2.1.8 Through plating hole

Although it is not a machine but a process, this technique is included in this section to exemplify this manufacturing process, which consists of a set of chemicals that can metallize holes without any machine but a vacuum bed and a reflow oven. There are

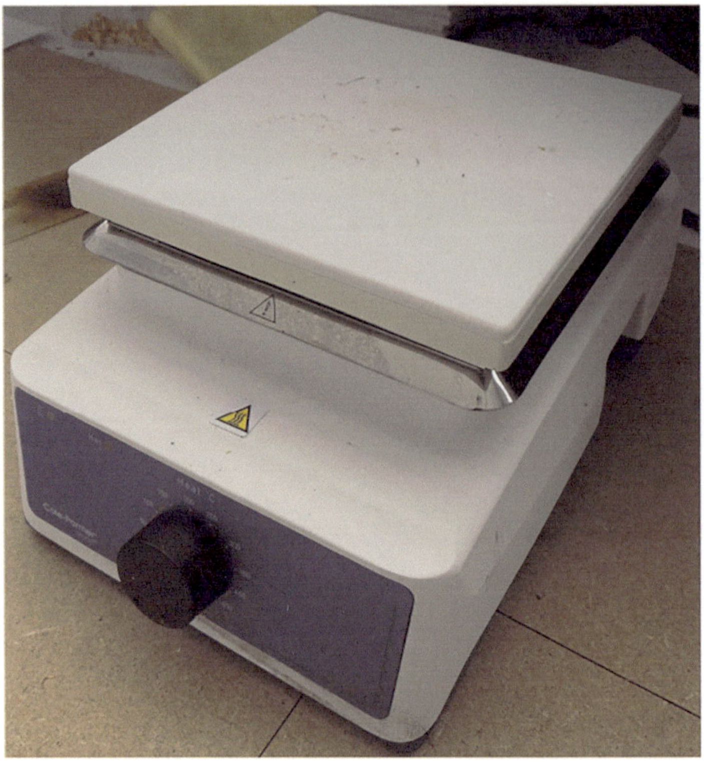

Figure 5.
Soldering plate.

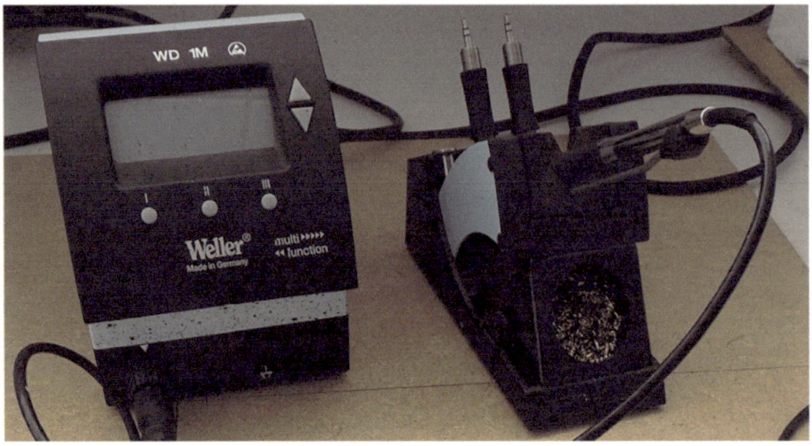

Figure 6.
Welder.

many commercial solutions in the market, and the results are good enough. The vacuum bed makes this dense liquid metal to cross along the wholes, and then, it must be treated with the reflow oven. Laser and milling machines usually include this vacuum bed in their systems. **Figure 8** shows an example of through plated holes.

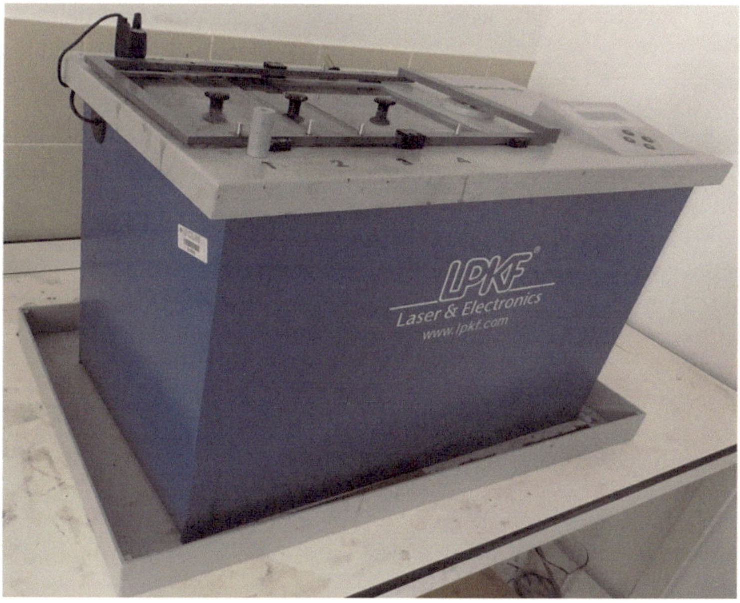

Figure 7.
Galvanic metallizer.

Figure 8.
Through plated holes.

2.2 Measure machines

In addition, high-performance manufacturing machines are needed; so essential are also the measuring machines to improve the results. When the design frequency is increased, the dimensions of the prototype are reduced and manufacturing tolerances become more important, being necessary to characterize and avoid them as much as possible. To do that, some of the measuring machines explained next could be used.

2.2.1 Microscope

It is very important to check the correct metallization of holes and walls, the correct milling of the prototype to avoid shortcuts, etc. To do that, a microscope can be used to check if there are manufacturing errors such as parts of copper that have not been properly removed or vias that are not properly metallized. Once identified, some of these errors can be handly solved to assure the correct behavior of the prototype.

2.2.2 Vision measurement system

This kind of machines are high-efficient systems that allow very high accurate measurements. These machines measure not only distances, but also geometrical 2D shapes, such as holes or lines and their deviations in shape and size from the initial requirements, allowing to compensate systematic errors during the manufacturing process.

2.2.3 Profilometer/surface roughness measuring systems

These instruments measure the rugosity of the layers of the prototypes. They are very useful in order to estimate the effect of metallization.

2.2.4 Vector network analyzer

These N-port devices are capable of measuring S-parameters, magnitude, phase, and time-domain analysis of prototypes. Combined with an anechoic chamber, it can measure antennas and radiation of prototypes.

3. Layer manufacture

As it was explained formerly, to create a 3D structure, an undetermined number of layers have to be piled up. These layers can have different heights and can be manufactured using the same or different substrates, with a careful alignment of them being necessary. **Figure 9** shows an exploded schematic view of a prototype. It shows different layers with different geometries and different heights making up the whole 3D structure.

Figure 10 shows a full assembled prototype besides a euro coin to compare sizes, where it is possible to observe how different layers with different heights (among 0.5 mm and 0.8 mm) are assembled.

These layers are manufactured using standard PCB processes such as cutting, drilling, milling, and plating. Once the layers are manufactured in the PCB, they are carefully extracted to be piled up. **Figure 11** shows the prototype layers before being extracted from the PCB. These PCBs are slabs that can be none, single, or twice metallized. Metallized substrates present a thin metal layer in one or both faces, usually copper. Later on, these PCBs can be additionally metallized such as the slab at the right in **Figure 11** using the process explained in section 2.1.7.

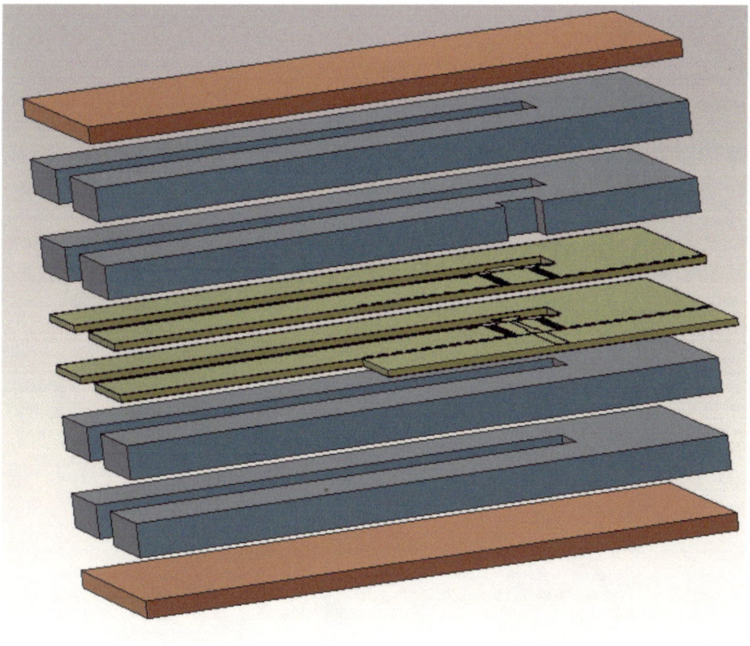

Figure 9.
Exploded view of the layers of a prototype.

Figure 10.
Assembled prototype.

4. Set-up issues for measurement

In this section, some example of feeding lines to feed the prototypes and connect them to a vector network analyzer are shown. Furthermore, a detailed explanation of each one of the issues that can appear during the manufacturing process is expound and analyzed.

4.1 Feeding lines

3D structures based on planar layers must include planar feeding lines such as microstrip lines or coplanar-grounded waveguides to feed the prototypes and connect them to a vector network analyzer. Moreover, prototypes fed with these lines can be accurately calibrated with TLR (Through-Reflect-Line) calibration kits. That allows to move the reference plane to the beginning of the waveguide and removes the losses and effects of the feeding line and connectors. It is recommended to design the prototypes with the same length as the calibration kits, so the same kit can be used to measure multiple prototypes. For different frequency bands, feeding lines with different lengths and widths must be used, and obviously, different calibration kits will be required.

Figure 11.
Layers in PCB.

Depending on the connectors used to measure, the design of the feeding line changes slightly. Usually, these connectors are either typical removable coaxial RF connectors or soldered ones. In **Figure 12**, some calibration kits can be seen with different connectors.

4.2 External metallization

The main raw material for building 3D structures piling up layers is PCBs; there are lots of commercial choices with different characteristics and heights depending on the working frequency and requirements of the prototype. These substrates are made of different components and have none, one (top or bottom), or two (top and bottom) thin layers of copper or other metals. The main idea is to stack different layers of different heights extracted from these substrates to create a whole 3D structure.

With the aim of building a metallic external wall, some layers must be externally metallized, and this can be done with a die cut system similar to the example shown in **Figure 13**. As it can be seen, the layer is almost fully metallized externally except for the small tabs located in the corners, which are needed to hold the structure to the PCB. The substrate around the external metallization of the prototype can be removed by simple exterior cuts shortened enough to hold the structure. The result is a completely metallized edge of these layers, but for the small corners as it was noted before.

4.3 Machine corrections

To reduce errors during the manufacturing process, machinery side effects must be identified and characterized. In case of using a multipurpose UV Laser System such as a cutter/plotter laser, these lasers are slightly tilted [11]. So that substrates will be

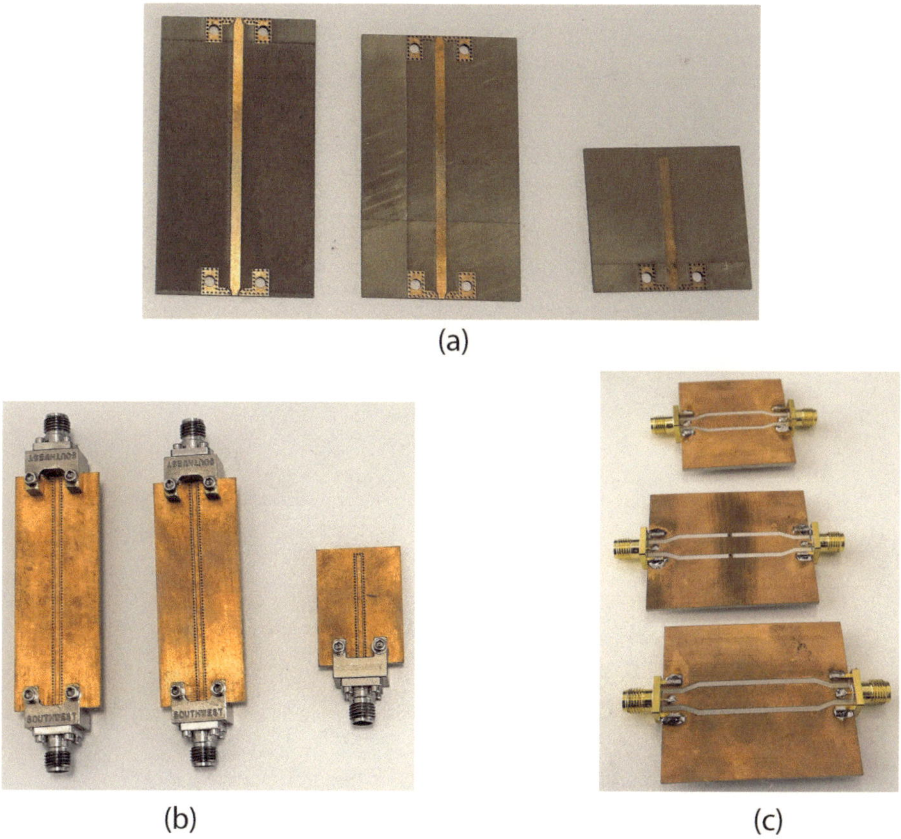

Figure 12.
Calibration kits. a) Microstrip calibration kit without removable connectors, (b) coplanar calibration kit with removable connectors, (c) coplanar calibration kit with soldered connectors.

cut with certain angle, producing a small ramp instead of a 90° cut (see **Figure 14**), which can be measured with a microscope or a vision system to incorporate the proper correction during the design phase of the prototype. Thus, an over-cut equal to the measured deviation compensates the effect of non-perpendicular laser cutting.

As it is shown in **Figure 15**, this ramp can be divided in three points: 1 (upper point of the ramp), 2 (lower point of the ramp), and 3 (middle point or average point of the ramp). Without correction or overcut, points 2 and 3 are lightly smaller than they should be. Over-cutting enlarges this cut adding this laser error and moving the desired size to point 3, the middle point, instead of point 1. This shifting produces point 1 to be a bit larger and point 2 to be a bit smaller; however, on average this cut is just as it should be. By doing this, the frequency shift presented by selective devices, such as filters, is drastically reduced, with the measured parameters of the prototype being more similar to the simulated ones.

Some cuts have complex geometries or are very tiny. These machines may have problems doing them correctly, because of that, additional cutting phases with simple shapes in certain areas are interesting to solve these issues. The same can happen for difficult milling areas and the solution is similar; additional milling layers should be defined, but it must be done carefully to avoid substrate from burning or getting damaged.

Figure 13.
Die cut example.

Although these machines work similar along the time, tools may need to be adjusted manually to seek for optimal performance every few months. This can be done easily using scraps of previous substrates.

If the prototype has holes used to pass through a wire or alignment pins, the overcut must be double to assure enough clearance in the whole hole to pass the wire or pin properly. In case there is a metallization, holes and cuts are reduced due to the presence of the additional metal and need to be taken into account to define the needed overcutting. Metallization can be either measured or calculated depending on the process, with the measured value being preferable when possible.

In case of using a mechanical cutter such as a drilling machine, the drill is really close to a perfect perpendicular cut. However, drilling tools are usually bigger or smaller to the theoretical size, due to fabrication tolerances. To avoid this, the size can be easily measured by doing a small cut in the substrate that we are using and measuring it, applying the proper correction in case it was needed.

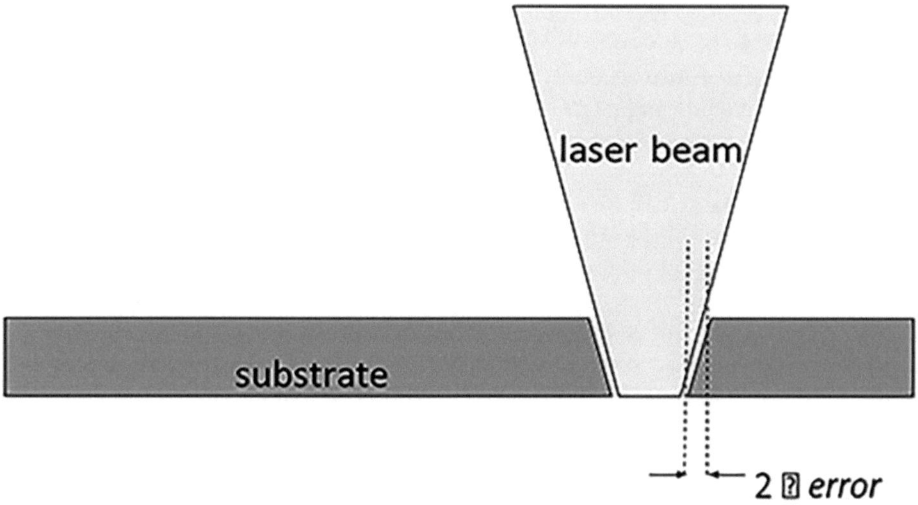

Figure 14.
Laser cut error. Lasers are slightly tilted, and this produces a small ramp instead of a 90° cut with some negative impact.

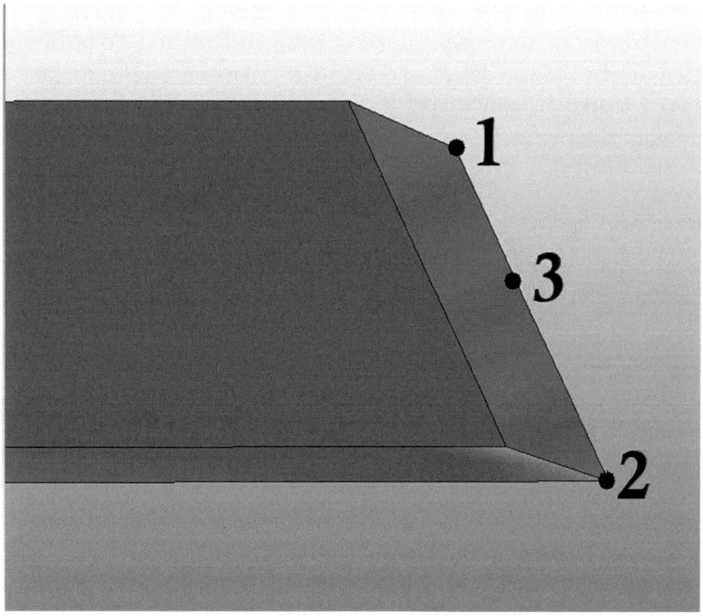

Figure 15.
Laser cut error, Figure extracted from [12].

The absence of these corrections can produce frequency shiftings and misfunctions in the performance of the prototypes, among other effects.

4.4 Assembling and weldering

One of the easiest and better ways to ensemble properly the layers of the prototype is by using alignment rivets, screws, or dowel pins. The position of the alignment holes to pass the pins through is arbitrary as long as they are outside the sensible areas of the circuit, but need to be the same in all the circuit layers [13].

Taking the former into account, the process consists of stacking all the circuit layers and roll out soldering paste between layers. Once all the layers have been stacked, a heating process must be done to melt the paste and weld the different layers of the prototype.

The soldering paste can be rolled out manually, but an automatic dispenser will provide more repeatable results. The soldering paste must be set by small dots and evenly separated among them. If there is too much paste, it will overflow inside the cavity, increasing the losses of the prototype, and producing missmatches and undesired frequency shifts. On the other hand, if there is not enough soldering paste, this will produce a poorly welded device with a highly degraded performance. After having manufactured several prototypes, the technician will have the experience to know almost the exact amount of paste to be used for each kind of circuit.

The first option to align the layers could be to use alignment screws with nuts properly tightened to make the needed pressure all around the prototype and to weld the soldering paste in a reflow oven. However, pressure could be not evenly distributed on the welding area, as it depends on the number and distribution of the screws. Moreover, the pressure done by the screws directly over the prototype layer can externally damage the structure. To prevent this, press covers can be used.

Press covers consist of some additional non-metallized and non-adherent layers that are located above the top cover and below the bottom one to make a more uniform pressure along the prototype. If covers were metallized or adherented, they could be welded to the prototype due to an eventual overflow of soldering paste. Once the prototype has been properly soldered, press covers should be removed, as well as screws and dowel pins.

When the welding among layers is not good enough, the discontinuities between successive layers can result in small cavities. This effect, for structures such as ESIW, will cause the magnetic currents to modify their path by coinciding these cavities with their direction due to the TEM mode (Transverse Electro-Magnetic), degrading the response of the device and causing anomalies. In ESICL lines, however, the magnetic currents do not coincide with the possible discontinuities (cavities) produced; for this reason, welding is not necessary in these prototypes, and screws and nuts can assure a proper performance of the device, this assembling process being quicker due to the lack of welding process.

Rivets work similar to screws for alignment, being the pressure along the prototype uneven and not as tight as screws.

Dowel pins combined with a multipress usually produce better results as pressing machines press evenly the circuit. For this process, press covers (additional layers similar to those described to be used with screws) must be used to protect the circuit. If pressing machines do too much pressure on the circuit, this can lead to a bending of the covers deforming the inner cavities. To prevent this, different press covers adapted to the geometry of the welded area must be designed [14]. The same idea improves the results using screws and a reflow oven; however, whenever possible, press machines and dowel pins give much better results than screws and reflow oven.

Another thing to be taken into account is that the presence of soldering paste produces a little layer of a few microns that can enlarge/enheight the waveguide, with enough pressure while the paste melts this layer being reduced, and this effect can be negligible.

4.4.1 Manufacturing process of vias

Although soldering and fencing vias have different purposes, from the manufacturing point of view, both are metallized holes along the surface layers.

Soldering vias: Soldering vias are metallized holes used to fill in with soldering paste instead of spreading along the layers or to be used as a leak for excessive soldering paste.

Fencing vias: To assure a good isolation of the prototype, sometimes it is necessary to use fencing vias that work as a perfect electromagnetic wall [15], and whose diameter changes with the frequency at which the prototype is designed and the technology used. Fencing vias must be metallized; otherwise, the prototype will have big losses due to the resonant effect of these holes.

These metallized elements can work as resonators if the wavelength of the prototype is similar to the size of these vias, the same happens with the separation between them. The vias must follow these rules and use the formulas for SIWs design [15]:

- They must be equally separated in each row.
- They must be equally separated from the edges of the prototype.
- They must be small enough compared with the wavelength of the prototype.
- They must be separated enough compared with the wavelength of the prototype.

Figures 16 and **17** show an schematic view of fencing vias located around cavities. The layer in **Figure 16** has holes for alignment purposes (blue), vias (purple), and the inner cavity (yellow) named as *long*. In the example of **Figure 17** a separation of 0.1 mm toward the cavity was chosen to fulfill the requirement explained above.

4.5 Procedures requiring more than one soldering phase

Some prototypes, for example, RESIW or DRESIW devices [16], may need to be welded in two or more phases. To make possible a welding process divided into two or more phases, the methods described previously can be combined together. The only requirement for multiple soldering phases is to use different solder pastes with different melting temperatures, so that the following phases will not de-solder the already soldered layers in previous ones. For that, the soldering paste with the highest melting temperature will be used for the initial phase, and soldering pastes with lower melting temperatures will be used afterward when these phases contain the already welded parts of the circuit. This procedure is very useful for ridge prototypes; ridge prototypes may need different pressing covers with different geometries of each one of the pressing phases. **Figure 18** shows these press covers.

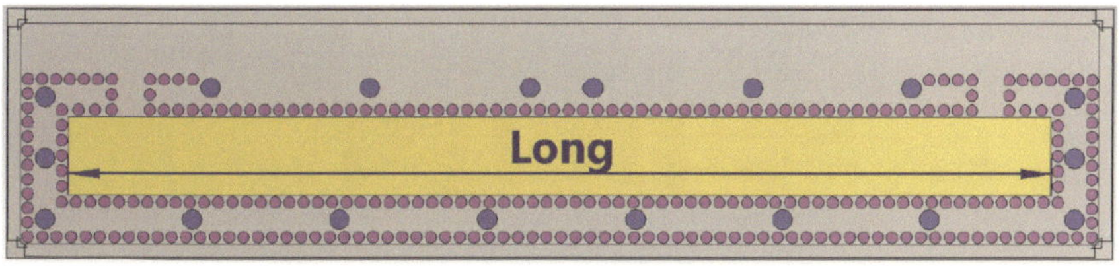

Figure 16.
Vias position.

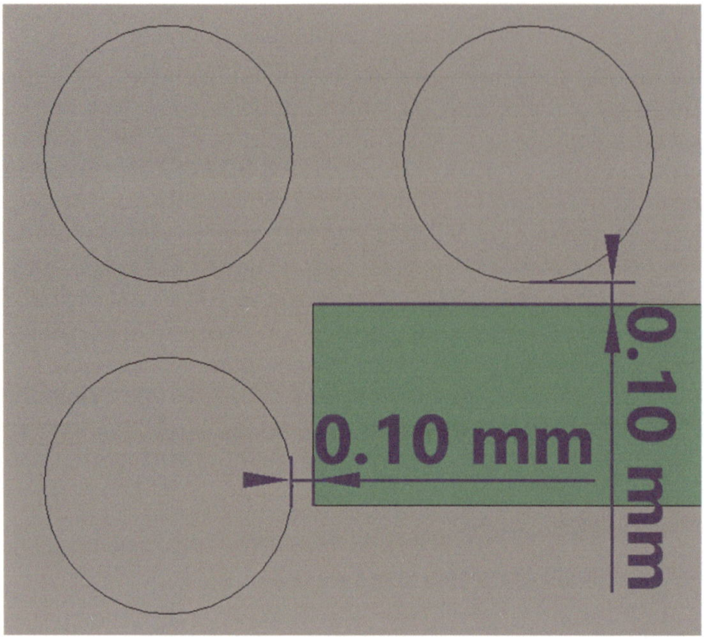

Figure 17.
Vias position toward edge example.

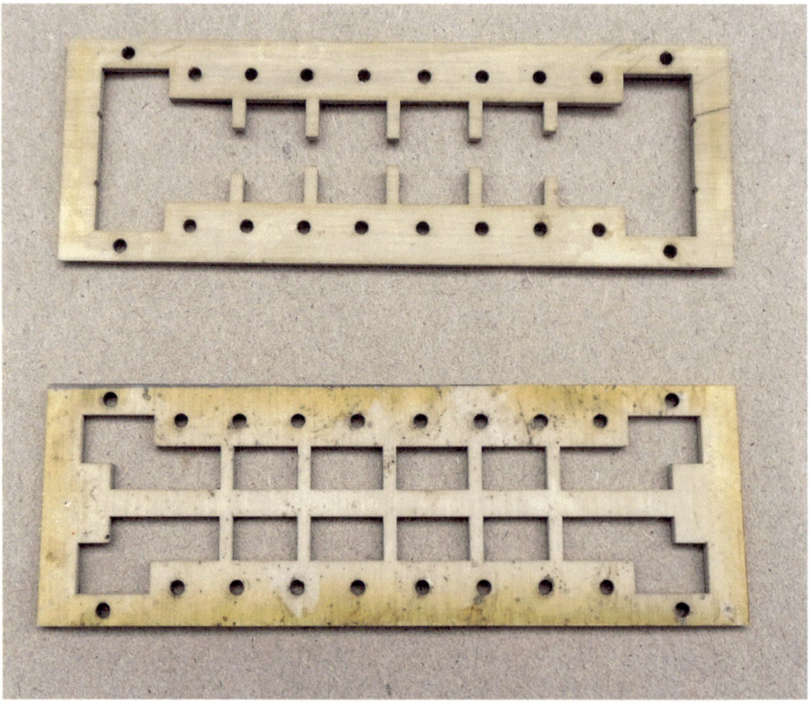

Figure 18.
Press covers of an RESIW filter [14].

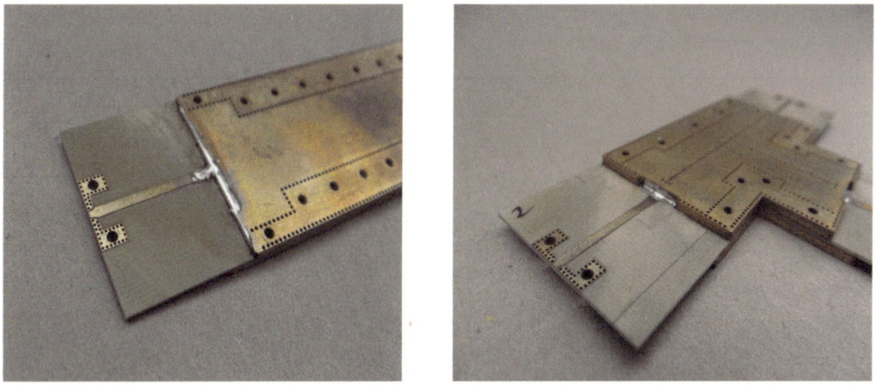

Figure 19.
Handly backing weld prototype examples.

This technique is also quite useful when prototype junctions are not good enough and need to be fixed manually with soldering paste and a welder. Moreover, if there was an overflow of solder paste, it can be removed with a de-soldering iron or roll out manually with a welder. These press covers are different to apply differently pressure in each soldering phase [14].

4.6 Handly backing weld

Some prototypes have sensible areas that need to be welded once the whole circuit has been ensemble. The idea is to roll out some solder paste in this places and melt it with a welder to ensure a good welding. In **Figure 19**, examples of this can be seen.

4.7 External elements

External elements, such as wires and rods, are very useful in order to design transitions [10, 17] or filters [18]. In the case of transitions, wires are used to connect the layers of the prototype (rods can also be used if they are small enough). On the other hand, rods are used as filter resonators.

Two factors are essential when wires and rods are used in the prototypes: their position and shape. Wires must be stretched as much as possible in order to reduce deformations and foldings of itself. On the other hand, the easiest way to ensure the position is through a hole. To assure a good electrical contact in the weldering, the hole must be fully metallized, a galvanic or a through-hole plating metallization is recommended for that. The metallization must be taking into account, as the hole will narrow some microns. If a laser cutting machine is used, the diameter of this hole must be found out by testing different diameters in order to take into account the laser error due to the non-perpendicular hole and the reduction of the hole due to the metallization process, as it was explained in Section 4.3. **Figure 20** shows an example of this testing; in each column each hole is increased by a few microns regarding the previous one.

After the wire is crossed through the hole, it has to be welded to the hole with soldering paste. It must be done trying to minimize the residues of this paste. Once

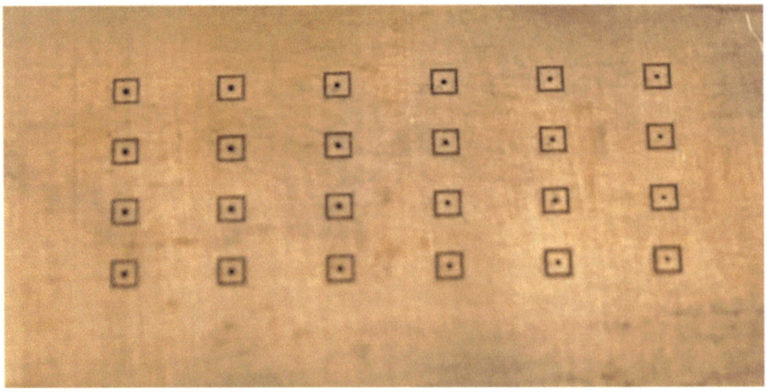

Figure 20.
Swept hole diameter.

this wire is welded, the spare wire must be removed by cutting it. Then, depending on the structure of the prototype, the other end of the wire can be welded to its own hole. While crossing the wire through the whole, if the hole is very tight, metallization can be detached as the wire pushes it away. Therefore, electrical contact would be spoiled, so that it must be done carefully. If metallization is taken away, performance of the prototype will degrade.

There are other considerations to take into account. If there are more welding phases after weldering the wire, a soldering paste with higher melting point must be used to weld the wire to avoid de-soldering in the following processes which will have lower melting points.

The higher the conductivity of the wire is, the better the results will be. However, due to the small size of these elements in microwave devices, conductors such as brass or tinned copper are enough to achieve good results.

When possible, wires should be replaced by rods. In case wires were used, they must be stretched in order to reduce foldings and deformations.

5. Conclusions

3D prototypes can be done with planar procedures piling up different layers to build three-dimensional structures. The techniques used for manufacturing planar circuits can be adapted to build these prototypes, which can be combined with external elements such as wires or rods.

Different modifications have been shown to improve results in each manufacturing phase; errors and specific features of each machine must be taking into account to reduce its impact in the manufactured layers. Over-cut, under-cut, measurement and estimations are some of the techniques used to reduce these effects.

Regarding the assembling techniques with screws or dowel pins, the issues due to pressure, excess/lack of soldering paste, etc., should be considered. Finally, the use of press covers is also recommended in soldering phases, taking into account that technologies such as RESIW or DRESIW have some variations.

All the information collected in this chapter has been extracted from empirical evidence and real circuits that have been manufactured.

Acknowledgements

This work was supported by the Ministerial de Cuenca e Innovación, Spanish Government, through the Subproject C44 of the Coordinated Research and Development Project PID2019-103982RB under Grant MCIN/AEI/10.13039/501100011033.

Thanks to Gamma research group (Grupo de Aplicaciones de Microondas y Milimétricas, y Antenas), url: https://gamma.uclm.es/.

Author details

Darío Herraiz[1*†], Leticia Martínez[2†], José A. Ballesteros[1†], Marcos D. Fernandez[1†], Héctor Esteban[3†] and Ángel Belenguer[1†]

1 Escuela Politécnica de Cuenca (EPC), Cuenca, Spain

2 Instituto de Tecnología, Construcción y Telecomunicaciones (ITct), Universidad de Castilla-La Mancha Campus Universitario, Cuenca, Spain

3 Escuela Técnica Superior de Ingeniería de Telecomunicación, Universitat Politècnica de València, Valencia, Spain

*Address all correspondence to: dario.herraiztirado@uclm.es

† These authors contributed equally.

Contratado predoctoral UCLM, financed by European Social Fund +(ESF+).

IntechOpen

© 2022 The Author(s). Licensee IntechOpen. This chapter is distributed under the terms of the Creative Commons Attribution License (http://creativecommons.org/licenses/by/3.0), which permits unrestricted use, distribution, and reproduction in any medium, provided the original work is properly cited.

References

[1] Deslandes D, Wu K. Integrated microstrip and rectangular waveguide in planar form. IEEE Microwave and Wireless Components Letters. 2001; **11**(2):68-70

[2] Belenguer A, Esteban H, Boria VE. Novel empty substrate integrated waveguide for high-performance microwave integrated circuits. IEEE Transactions on Microwave Theory and Techniques. 2014;**62**(4):832-839

[3] Herraiz D, Esteban H, Martínez JA, Belenguer A, Boria V. Microstrip to ridge empty substrate-integrated waveguide transition for broadband microwave applications. IEEE Microwave and Wireless Components Letters. 2020; **30**(3):257-260

[4] Herraiz D, Esteban H, Herraiz D, Vidal A, Belenguer A, Boria VE. Microstrip to double ridge empty substrate integrated waveguide transitions based on exponential and superelliptical dielectric taper. IEEE Access. 2021;**9**:165745-165753

[5] Gatti F, Bozzi M, Perregrini L, Wu K, Bosisio RG. A novel substrate integrated coaxial line (SICL) for wide-band applications. In: 2006 European Microwave Conference (IEEE). 2006. pp. 1614-1617. DOI: 10.1109/EUMC.2006.281409

[6] Belenguer A, Borja AL, Esteban H, Boria VE. High-performance coplanar waveguide to empty substrate integrated coaxial line transition. IEEE Transactions on Microwave Theory and Techniques. 2015;**63**(12):4027-4034

[7] Borja AL, Belenguer A, Esteban González H, Boria VE. Design procedure of continuous profile stopband filters implemented with empty substrate integrated coaxial lines. IEEE Transactions on Microwave Theory and Techniques. 2020;**68**(4):1520-1528

[8] Juan A, Belenguer A, De Dios JJ, Esteban H, Boria V. Wideband transition for increased-height empty substrate integrated waveguide. IEEE Access. 2019;**2019**:1

[9] Quiles F, Belenguer Á, Martínez JÁ, Nova V, Esteban H, Boria V. Compact microstrip to empty substrate-integrated coaxial line transition. IEEE Microwave and Wireless Components Letters. 2018; **28**(12):1080-1082

[10] Belenguer A, Ballesteros JA, Berlanga MF, Esteban H, Boria V. Versatile, error-tolerant, and easy to manufacture through-wire microstrip-to-ESIW transition. IEEE Transactions on Microwave Theory and Techniques. 2020;**2020**:1

[11] Martínez JA, Belenguer Á, Borja Hectór Esteban AL, Boria VE. Corrección de errores en la fabricación de dispositivos en guía vacía integrada en substrato (ESIW) y línea coaxial vacía integrada en substrato (ESICL) [Error correction for manufacturing Empty Substrate Integrated Waveguide (ESIW) and Empty Substrate Integrated Coaxial Line (ESICL) devices]. In: XXXIII National URSI Conference in Universidad de Granada. Granada; 2018

[12] Tirado DH. Design, manufacture and behaviour of a microstrip-ESIW-E transition in Ka band [Final degree project]. Escuela Politécnica de Cuenca (EPC) – Universidad de Castilla La Mancha (UCLM), 2019

[13] Martinez JA, Belenguer A, Esteban González H. Highly reliable and

repeatable soldering technique for assembling empty substrate integrated waveguide devices. IEEE Transactions on Components, Packaging and Manufacturing Technology. 2019;**9**(11): 2276-2281

[14] Herraiz D, Esteban H, Morro JV, Herraiz D, Belenguer A, Boria V. Bandpass filter in ridge empty susbtrate integrated waveguide with U-Shaped impedance inverters. IEEE Access. 2019; **9**:165745-165753

[15] Xu F, Wu K. Guided-wave and leakage characteristics of substrate integrated waveguide. IEEE Transactions on Microwave Theory and Techniques. 2005;**53**(1):66-73

[16] Herraiz D, Esteban H, Martínez JA, Belenguer A, Cogollos S, Nova V, et al. Transition from microstrip line to ridge empty substrate integrated waveguide based on the equations of the superellipse. Applied Science. 2020;**10**: 8101

[17] Fernandez MD, Ballesteros JA, Belenguer A. Highly compact through-wire microstrip to empty substrate integrated coaxial line transition. Applied Sciences. 2021;**11**(15):6885

[18] Casero I, Ballesteros JA, Fernandez MD, Herraiz D, Belenguer A. Easy-to-assemble and high quality-factor ESIW filter with post-based soldered inverters in X-band. AEU—International Journal of Electronics and Communications. 2021;**142**:153987

Chapter 6

Metal 3D-Printing of Waveguide Components and Antennas: Guidelines and New Perspectives

María García-Vigueras, Lucas Polo-Lopez, Charalampos Stoumpos, Aurélie Dorlé, Carlos Molero and Raphaël Gillard

Abstract

This chapter intends to show the strong potential brought by metal 3D-printing to the field of waveguide components and antennas. General co-design guidelines are firstly provided. These guidelines enable to benefit from the advantages associated to metal 3D-printing. The implementation of filters and ortho-mode transducers is considered, together with horns and slotted antennas. Finally, multifunctional periodic structures benefiting from metal 3D-printing are discussed.

Keywords: 3D-printing, additive manufacturing, selective laser melting, ortho-mode transducers, filters, polarizers, leaky-wave antennas

1. Introduction

3D-printing has made a profound impact in the development of radio-frequency (RF) devices in the past decade. Initially, this manufacturing technique represented an opportunity to implement new ideas as well as to enjoy fast, cost-effective, and monolithic prototyping [1–8]. However, non-negligible constraints were found in the first attempts to 3D-print RF components. Such constraints were associated, for example, to surfaces rough finish, the appearance of undesired supports, and the need for electroless plating. Neither the resulting manufacturing tolerances were optimal, nor the quality of the printed pieces. With time, as a result of collective efforts, it has been understood that the full benefit of this technique (i.e., short lead time, low cost, single piece prototyping, and high RF quality) implies a change of mindset in RF design [9–13]. As a result, today we are witnessing the most disruptive impact of 3D-printing: the challenge of the bounds of our creativity and the need for new co-design guidelines.

3D-printing consists in the layer-by-layer additive manufacturing (AM) of objects. Different materials can be employed to build the piece. For example, plastic polymers can be used to build objects through stereolithography (SLA) [6, 7]. The case of the present chapter considers selective laser melting (SLM), which allows to use metal alloys as the building material [10, 14–16]. Examples found in the literature include aluminum (which is the most extended one, AlSi10Mg), titanium, stainless steel, or Invar. This option is the preferred one when dealing with aerospace applications. The main associated advantage is that it directly leads to a body that can easily meet the stringent mechanical and thermal conditions associated to space or other harsh environments. Additionally, metallic 3D-printed parts are compatible with high-power handling, specially if they are monolithic, as passive inter-modulation issues are minimized.

The main disadvantage of metal 3D-printing is the surface finish of the 3D-printed part. The resulting surface roughness is larger than what is normally obtained with conventional manufacturing techniques, which results in higher insertion losses. Consequently, 3D-printed parts might exhibit higher loss than the counterpart produced with CNC milling. For that reason, postprocessing and additional surface metallization techniques are currently being developed in order to improve the conductivity of a metal part [17]. The effective conductivity of a 3D-printed waveguide depends on many aspects such as the frequency of operation, the chosen printing direction, or the shape; however, an average value for a raw (not metallized) finish is 0.5×10^7 S/m. With treatments, values of up to 3×10^7 S/m can be achieved. This is still less than the ideal conductivity of aluminum; however, it has been noticed that the devices do give competitive loss thanks to the avoidance of assembly, which might create leakage or reflections.

Current metal 3D-printers offer printing volumes that range from 250 mm × 250 mm × 250 mm [18] to 400 mm × 400 mm × 400 mm [19], which are well suited for devices operating from X-band. At the time this chapter is being written, one can find stories in the media where the main space primes have announced the use of 3D printing in their future high-throughput satellites [20–22]. On the other hand, in 2021, a European Cooperation for Space Standardization (ECSS) standard with requirements for processing and quality assurance of powder bed fusion technologies for space applications has been issued [23]. All this together proves that the space industry recognizes the maturity of metal 3D-printing and accepts its use for present and future systems.

The authors of this chapter have been involved in several R&D projects related to 3D-printing of RF components and antennas. Such projects have been mainly conducted in France, at the laboratory IETR (Institut d'Electronique et des Technologies du numéRique), and they have been funded by the European Union, the European Space Agency (ESA), the Centre National d'Etudes Spatiales (CNES), and the Region of Brittany. Such projects have allowed the conception of general design guidelines for successful SLM prototyping, which are next provided in Section 2. In the following sections, advanced 3D-printed parts are described. Section 3 includes the implementation of waveguide components, such as filters and ortho-mode transducers. Section 4 considers the development of horns and slotted antennas. The case of periodic structures implementing frequency selective filtering and polarization conversion is considered in Section 5. Finally, conclusions are drawn in Section 6.

2. General design guidelines

A general belief when approaching 3D-printing is that any RF geometry can be fabricated. Still, while there is significant freedom and possibility to fabricate complex shapes, there are also a series of rules and guidelines that allow to obtain optimal results out of 3D-printing.

The set of guidelines is indeed very broad, since it goes from the initial RF design of geometry to the type of machine parameters used to fabricate the part. This section will only cover the ones concerning the RF designer, since the rest will change depending on many considerations, ranging from the type of machine to productivity. In any case, and as it will be demonstrated in this chapter (which contains many examples from different authors and therefore different manufacturers), the basic RF guidelines are enough to produce very good performance regardless of the 3D printer, alloy, or manufacturer.

Concerning RF design, the most basic consideration when 3D printing RF devices is the orientation that the piece will have in the 3D printer. Such orientation determines the manufacturing tolerances, the degree of symmetry of the printed component, the eventual need of supports if the printer finds surfaces that are hanging, and the number of parts that one can print in the same platform. In order to attain both high precision and high symmetry, the piece should be oriented in a way that either the main waveguide propagation axis is parallel to the building direction (vertical printing) or the main waveguide propagation axis is perpendicular to the building direction (horizontal printing).

The remaining of the subsection discusses the basic guidelines that a designer may follow to adapt a waveguide device to vertical or horizontal printing.

2.1 Overhanging faces

Overhanging faces are those that are perpendicular to the printing direction and facing down, as illustrated in **Figure 1(a)**. As part of the design process, such overhanging faces should be modified, since they can suffer strong deformation or even collapse if they are too big. The recommended design practice is to chamfer or tilt the face with an angle of 45° or larger. As it will be shown later, this technique can be applied to corrugations, irises, cavities, posts, etc.

2.2 Wall thickness

In certain microwave devices, thin walls are needed as part of the RF design (e.g., septums in orthomode transducers, irises in filters, inner walls in power dividers, etc). To the best knowledge of these authors, the lowest wall thickness that can be 3D-printed nowadays with a good reliability is 0.5 mm. In certain

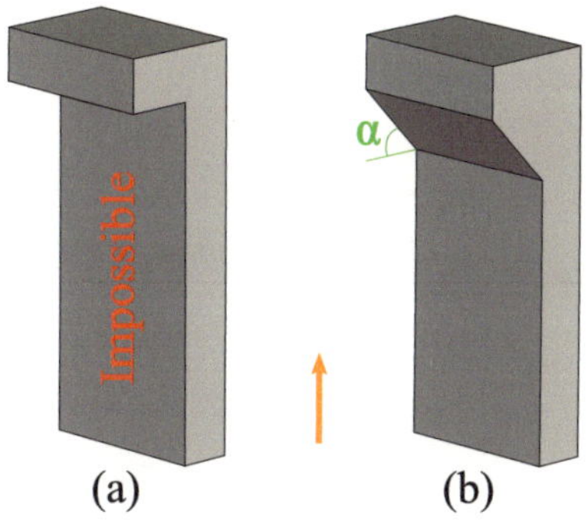

Figure 1.
Nonfeasible overhanging face (a) and feasible inclined overhanging face (b). The orange arrow indicates the printing direction.

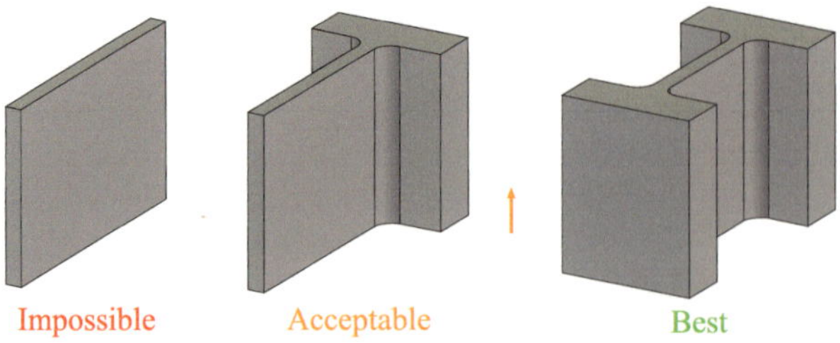

Figure 2.
In order to successfully print very thin walls, it is important that they are supported by adjacent thicker walls. The orange arrow indicates the printing direction.

occasions, it can be possible to achieve thicknesses of 0.3 mm, although only for very short walls and supported by thicker adjacent walls, as illustrated in **Figure 2**.

2.3 Height

When creating elongated structures such as posts, it is important to take into account that their height should not be greater than five times the diameter of the base. If this requirement is not satisfied, the post can suffer deformations. In the case that it would be necessary to use a feature that does not satisfy the previous criteria, the recommendation is to use a truncated post (or a prism whose base is larger than the top face), as it is indicated in **Figure 3**.

2.4 Rounded edges

Rounding of the edges of the structure is welcome for 3D-printing. The recommended rounding radius is at least 0.3 mm (although the larger, the better). The reason for this recommendation is the avoidance of local overheating when printing a certain edge, which can result in deformations of the structure [24].

2.5 De-powdering

The “de-powdering” is the activity where all the unmelted powder is removed from the fabricated part. The RF designer should conceive the device avoiding highly concave cavities (those whose only opening is very small when compared with the size of the cavity) as well as minimizing (if possible) the presence of siphon waveguide sections.

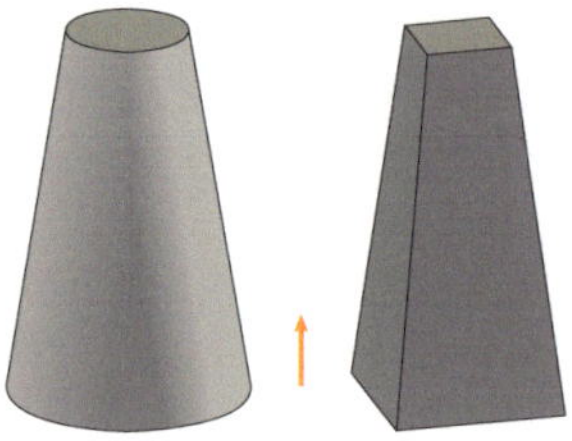

Figure 3.
Different approaches to increase the stability of elongated structures. The orange arrow indicates the printing direction.

2.6 Machining of waveguide flanges

Finally, a highly recommended RF practice is the reworking of the flanges using CNC milling in order to ensure a perfect connection with the rest of the devices.

3. Waveguide components

3.1 Filters

Waveguide filters are devices used to select and/or reject signals in an RF chain. The implementation in waveguide technology ensures low losses and high-power handling, while its main disadvantages are the associated mass and the volume. In contrast to other waveguide devices listed in this chapter, filters are, in general, rather sensitive to the manufacturing tolerances. As a result, the use of the previous design guidelines is even more relevant in the present case.

In the domain of metallic 3D-printed filters, it is possible to find multiple articles where AM is used to enable the prototyping of complex geometries [25]. In such examples, the printing direction has not been specifically considered as part of the design process, which is the purpose here. In the following, two examples are illustrated that are based either on vertical and horizontal printing (associated respectively to waveguides whose propagation axis is either parallel or perpendicular to the printing direction) (**Figure 4**).

In the case of vertical printing, traditional filter topologies shall be adapted by tilting all their down facing parts so that they become self supporting. A first example is proposed in [15], where $\lambda/4$ rejection stubs are tilted. The authors of such paper present prototypes in Ku/K-band that have been fabricated using different metal alloys. All the filters present high rejection (50 dB) and low insertion losses (from 1 to 0.1 dB, depending on the type of surface finish). Another example can be seen in [11], where a bandpass filter with titled irises is presented. The design in the article is a 11-pole filter operating from 17.3 to 20.2 GHz. The two fabricated filters exhibit a bandwidth slightly narrower than the simulated one. An interesting result is the great similarity between the two prototypes, which have been produced using different processes with the same alloy. The latter is a confirmation of what was stated in Section 1: using guidelines for 3D-printing enables robust filter designs. A third example can be found in [16], where a corrugated filter with chamfered corrugations is presented. The filter also operates in the Ka-band (17.7–20.2 GHz) and provides all-mode rejection up to 43 GHz. This work also provides a comparison between raw

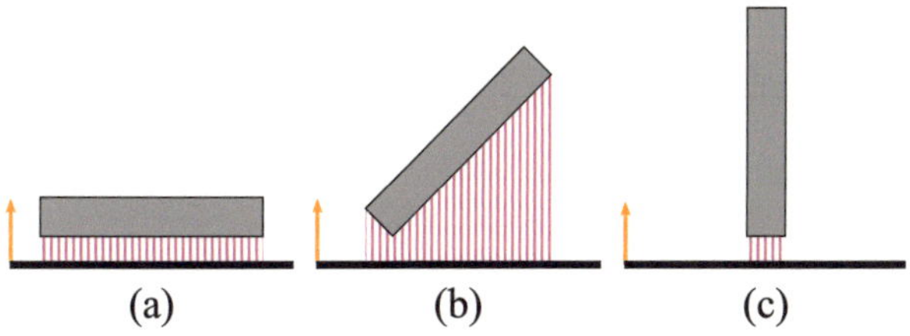

Figure 4.
Different orientations of a filter (gray) on the building platform (black) and the required supports (magenta) for each of them. The orange arrow indicates the printing direction.

(not metallized) and metallized finish, where the associated ohmic losses are 0.35 and 0.25 dB, respectively.

Interestingly, there are also filter examples that by default (without requiring dedicated design guidelines) are compatible with vertical printing such as the filters based on spherical or quasi-spherical resonators. In [26], a dual-mode filter at 8.25 GHz implemented with spherical cavities (which can support two modes) and compatible with vertical printing is presented. Dual-mode filters are known for being very sensitive and often require tuning screws; however, the 3D-printed filter in the article achieves a performance very close to simulation without the need of tuning screws, which is another demonstration of the suitability of vertical printing for the production of waveguide filters.

Ridge waveguide evanescent mode filters can also be adapted to vertical printing. These filters are based on the cascading of ridge and hollow waveguide sections of the same width and height. By chamfering the ridges as in **Figure 5**, the filter becomes vertically printable, as demonstrated in [12].

Figure 6 shows the measured and simulated performance of one filter of this kind. Two identical samples of the filter have been manufactured in order to verify its repeatability. As it can be appreciated, the results of both prototypes show excellent agreement between them and also with the simulation. The insertion losses (0.3 dB) are in agreement with the expected value for a nonmetallized component.

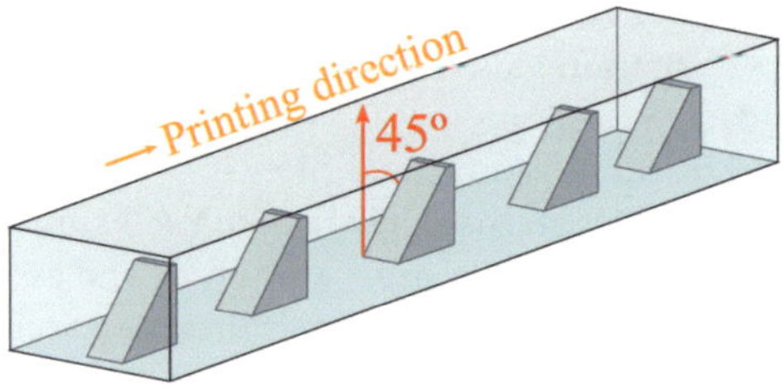

Figure 5.
Representation of an evanescent mode ridge waveguide filter designed to be vertically printable. The ridges have been chamfered with a 45° angle so that they become self-supporting.

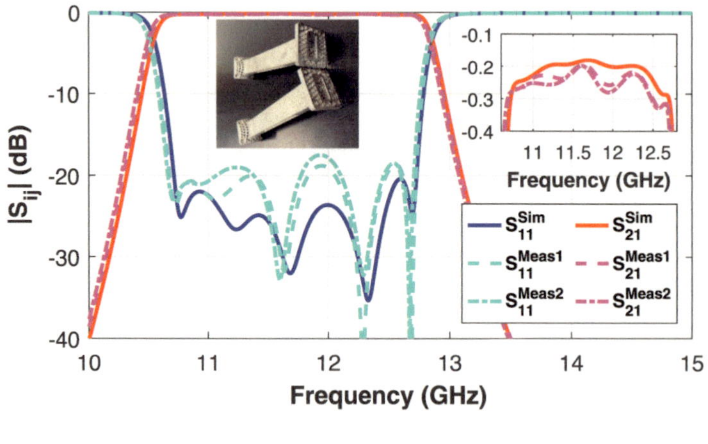

Figure 6.
Simulated and measured scattering parameters of an evanescent mode filter with chamfered ridges. Two copies of the prototype have been manufactured in order to assess the repeatability. These figures are shown here for the first time.

An example of horizontal printing can be found in [17]. In that work, the authors report a combline filter where the cavities have triangular cross section. When 3D-printing this filter, the combline post grows from the base of the triangle, while the other two sides of the triangle (with an angle with respect to the base larger than 45°) converge vertically until they find each other, thus closing the cavity.

Another common practice in 3D-printed filters is the introduction of features in order to improve the performance (quality factor enhancement, extension of spurious-free band, etc). One example can be seen in [11], where a dimple has been created at the center of each cavity. Such feature has negligible impact on the insertion loss but pushes up the repeat band.

3.2 Orthomode transducers and polarizers

Orthomode transducers (OMTs) are commonly used to separate or combine two orthogonally polarized signals from or into the same waveguide. OMTs typically operate in linear polarization, and its key figures of merit relate to the isolation between the rectangular ports and cross-polarization discrimination (XPD) between the signals in the dual-polarized port. Septum polarizers are a type of OMT that also converts from linear to circular polarization and vice versa.

The isolation and XPD of an OMT depend on the structural symmetry of the component [27]. Potential asymmetries as consequence of fabrication will produce unwanted coupling between signals that are supposed to be orthogonal. As it has already been mentioned, the structural symmetry is better assured when the components are printed in vertical direction, hence the importance of the design guidelines for OMTs.

It is worth starting by the contribution in [9], which reports both side-arm OMT and septum polarizer in the Ka-band that are adapted to vertical printing. While the septum polarizer does not require much effort for 3D-printing, the side-arm OMT has been redesigned in a way that the "side" waveguide presents the top faces tilted in order to be self-supporting. Several E-plane matching steps are required to obtain optimal matching. Both devices are single-band and exhibit measured isolation and XPD in excess of 30 dB.

In [28], the authors present a Ka-band dual-band Bøifot OMT also adapted for vertical printing. The back face of the inner septum in charge of splitting/combining the polarizations has been chamfered according to the printing direction. It exhibits isolation and XPD greater than 40 and 30 dB, respectively, over 18.6–20.2 GHz and 28–30 GHz.

Interestingly, and despite being one of the most commonly used OMTs, there is no work in the literature with vertically printable turnstile OMTs. Two design concepts are discussed here for the first time. The first concept of such a design is depicted in **Figure 7(a)-(b)**, which shows the CAD model of the five-port turnstile junction. The main difference with respect to conventional designs is the four rectangular single-polarized waveguides, which are tilted to the junction axis so that they do not show unsupported faces. As **Figure 7(c)** also proves, the performance of the modified junction (reflection coefficient level lower than −25 dB is achieved for the full Ku-band SATCOM bandwidth, 10.7–14.8 GHz) does not show any limitation despite the modification.

The second design concept, in **Figure 8**, exploits the use of truncated cones for the realization of the turnstile post. One design approach when the bandwidth of OMT gets larger is the use of multistep posts [29]. As it has been discussed earlier, this type of

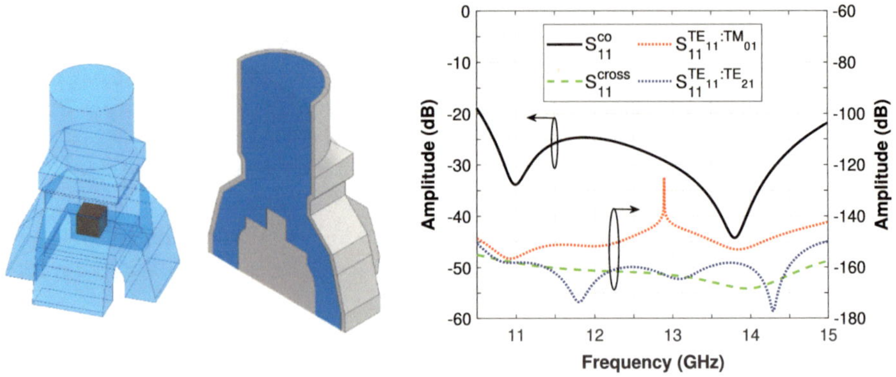

(a) (b) (c)

Figure 7.
First concept of a Ku-band (10.7–14.8 GHz) self-supporting turnstile-based OMT in vertical full-metal 3D-printing to avoid overhanging parts during the print process: (a) perspective view of the RF layout of the five-port junction (the transparent blue part is vacuum and the brown part is the metallic turnstile post), (b) transparent vertically cut view of the mechanical layout of the five-port junction (the blue part is vacuum and the gray part is metal), and (c) simulation results (port 1 refers to the common circular port).

feature is not well suited for vertical printing, hence truncated cones are used to design components with increased bandwidth (17.7–31 GHZ, with matching better than −20 dB in the uplink and downlink SATCOM bands). In this case, the OMT is terminated using E-plane power combiners designed in hexagonal waveguide. This second design has been fabricated and measured, and the results are depicted in **Figure 8(c)**. The measured reflection coefficients for both polarizations remain below −20 dB over the bands of interest (18–21.2 GHz and 27.5–31 GHz), while they also show a good agreement with simulations (not shown for clarity). The cross-polarized transmission coefficients and the rectangular ports present coupling levels below −35 dB and − 45 dB, respectively. The transmission coefficients are better than 0.5 dB (it must be noticed that the test setup requires extra waveguides in order to match the device nonstandard ports to standard ports and therefore the measured losses are higher than the ones of the component itself).

To complete the section, it is worth highlighting an interesting recent trend consisting in the combination of several OMTs in a single device to create a N-way OMT, possibility, which is enabled by the design freedom inherent to 3D-printing. The work in [30, 31] presents a four-way OMT-power divider, which is built as an array of 2 × 2 asymmetric side-arm OMTs in a tight square grid (around $1.25\lambda_0$) fed by two distributed 1 × 4 single polarized power dividers. The component was conceived without overhangs so it can be vertically printed. The measured prototypes feature isolation and XPD better than 40 dB.

4. Antennas

4.1 Horns

Horn antennas are widely used in various microwave and millimeter-wave applications, from feeds for reflectors to phased arrays or in antenna measurements. Such

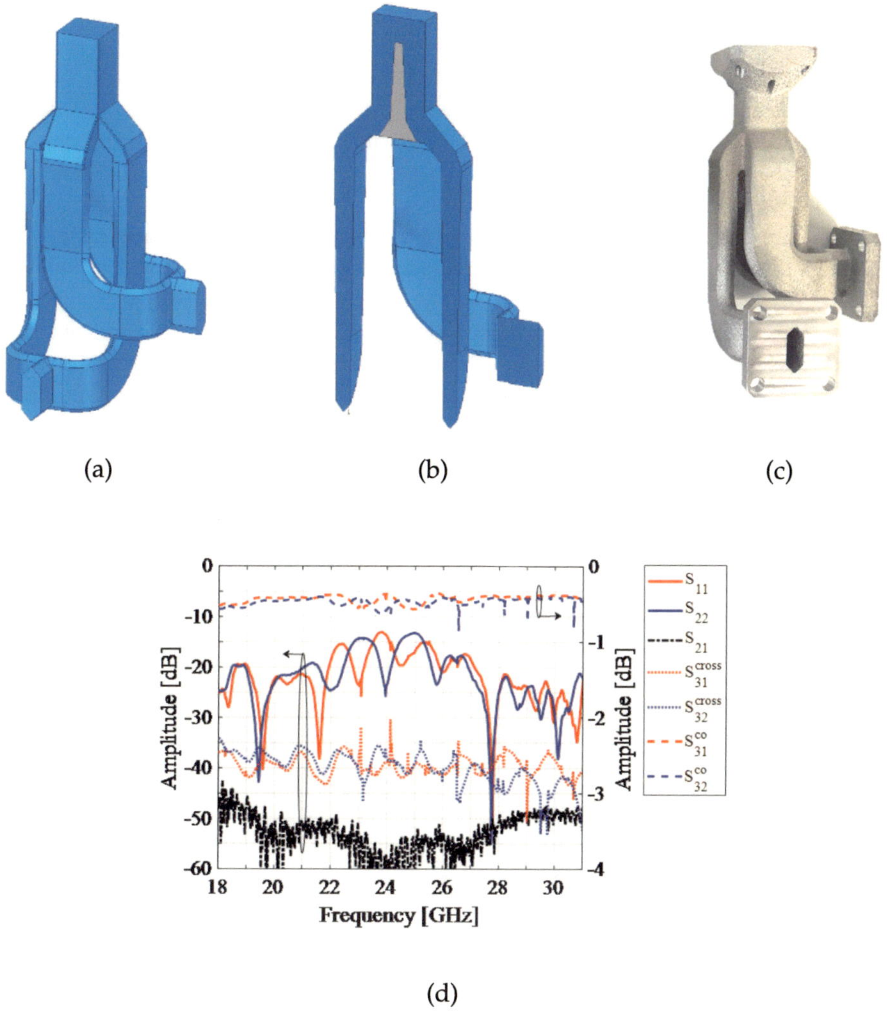

Figure 8.
Second concept of a dual-Ka-band (18–22 GHz and 27–31 GHz) self-supporting turnstile-based OMT in vertical full-metal 3D-printing to avoid overhanging parts during the print process: (a) perspective view of the total OMT's RF layout (the blue part is vacuum), (b) vertically cut view of the total OMT's RF layout (the blue part is vacuum and the gray part is the metallic turnstile post), (c) photograph of the printed prototype, and (d) measurement results. These results are shown here for the first time and they are courtesy of SWISSto12.

wide range of applicability is attributed to their robust RF performance such as medium or high gain, wide bandwidth, high XPD, and low losses.

As far as 3D-printing is concerned, horn antennas can be separated into two fundamental categories: (1) smooth-walled flared horns (this category includes different type of profiles such as pyramidal, spline, or Potter) [9, 31–36]; and (2) axially corrugated horns [37–39]. Regardless of the horn type, the printing direction shall be vertical in order to preserve the symmetries of the structure.

Stepped smooth-walled horns are discussed in [33], which presents designs operating in Ku-, Ku/K-, and Q/V-band. The three horns are printed vertically; consequently, the measured and simulated results of all antennas exhibit a good agreement. Particularly interesting is the results of the Q/V-band device, where a XPD better than −28 dB over the whole frequency band has been obtained experimentally.

Spline horns (smooth-wall horns whose flare follows a spline function) are presented in [32, 35]. These industrial works coincide in presenting Ku-band horns that are later integrated in a horn cluster (monolithic device including several horns as well as their corresponding feed devices). Moving on to horn clusters for the production of large horn arrays enables strong mass and cost reduction: on the one hand, the total number of parts (from bolts to mechanical brackets to the RF device) is drastically reduced, which has an impact on both the mass and the cost (dealing with all these parts requires an associated effort). On the other hand, the integration of a cluster is much simpler and faster, which reduces the overall program cost. This trend, which is enabled by 3D-printing, seems to be the future of Geostationary (GEO) telecommunication satellites [40].

In [34], a ridged horn antenna with multistep flare is presented. It is worth highlighting the high frequency achieved by this design (110 GHz), which is another proof of the very high performance that one could achieve when following the design guidelines. Moreover, extra features (corrugations) are added around the horn aperture in order for the horn to maintain high RF performance over a very wide bandwidth (45–110 GHz). This is another example of exploiting the design freedom of 3D-printing to improve the RF performance without increasing the manufacturing complexity.

Similarly, the work in [36] exploits the design freedom to consider perforations on the metallic walls of the horn to reduce mass without affecting the performance. The considered perforated gaps are smaller than $\lambda_0/15$ and hence opaque for the electric field. The experimental results demonstrate the suitability of such practice, which leads to a mass reduction in the order of up to two-thirds with respect to a horn.

To complete the survey of smooth-walled horns, it is also worth highlighting the work in [31], which presents a very high efficiency horn operating in the downlink Ku-band (10.7–12.75 GHz). The device, which is called quad-furcated profiled horn antenna, consists of four asymmetric small horns forming a 2×2 array that feeds a square waveguide aperture. The device also includes the feed network. All features in the component are compatible with vertical printing. The measured performance (S-parameters and radiation patterns) shows excellent agreement with the simulated one. This approach, which holds potential to use the power on board the GEO satellite more efficiently than other horn designs, can only be conceived with use of 3D-printing.

The design and fabrication of axially corrugated horns are presented in [37–39]. In particular, the accuracy and repeatability of 3D-printed choke horns at X/Ku band are investigated in [38]. Fifteen antennas were manufactured and tested, showing a coherent agreement between simulations and measurements in terms of matching, radiation efficiency, and radiation patterns. An axially corrugated horn covering the full Ka-band (26.5–40 GHz) is presented in [39], whose measured radiation patterns show good beam symmetry and XPD better than 29 dB. Reference [37] shows a choke horn that acts as feed in a transmitarray for cubesat applications. The antenna works in the range 23–26 GHz and is fabricated together with a septum polarizer.

Finally, to the best knowledge of these authors, there is no work reporting conventional corrugated horns where the corrugations are adapted to vertical printing. Nevertheless, the techniques described in [15] or [16] should be applicable to horns too.

4.2 Slotted antennas

Slotted waveguide antennas (SWAs) are attractive solutions for millimeter-wave applications because they enable high gain with a simple and flat beamforming network architecture. SWAs consist of a waveguide where one of the walls is periodically perforated with radiating slots. Generally speaking, SWAs can be grouped into two categories: resonant SWAs and non-resonant SWAs.

Resonant SWAs, which produce broadside radiation, use $\lambda_g/2$ inter-element spacing between the successive slots (where $\lambda_g/2$ is the wavelength in the waveguide) as well as they have a shorted termination. On the contrary, non-resonant SWAs do not produce broadside radiation and have inter-slot spacing that is different from $\lambda_g/2$ as well as a matched termination.

Regardless of the antenna type, SWAs exhibit a narrow bandwidth, which depends on several geometrical parameters, such as slot dimensions and shape, metal thickness, and inter-slot distance [41]. Nevertheless, this bandwidth can be widened using well-known design approaches such as separation of the array into several sub-arrays [42], use of ridge-waveguide [43], use of elliptical slots and direct coaxial feeding [44], as well as coupled-slot and differential feeding mechanisms [45].

Although the printing rules described in Section 1 apply also for slotted antennas, most of the works available in the literature report printing orientations of 45° with respect to broadside direction. Such works are not sensitive to the symmetry of the piece, therefore such orientation gives satisfactory results.

In [3], an 8x8-element resonant SWA in Ku-band with a corporate beamforming network is presented. Interestingly, this work also reports a counterpart based on subtractive manufacturing of several parts that are later assembled. The comparison between the two antennas shows that the 3D-printed antenna has a gain that is 1–1.5 dB larger than the traditionally manufactured antenna. In the latter, gaps and alignment errors between metal layers cause leakages and reflections, which are critical enough to eventually degrade the antenna performance. The article also discusses the fabrication orientation and the fabrication supports, which are the main disadvantages of this manufacturing approach. The same authors expand their work on SWAs in [46], which also operates in Ku-band but is much larger (16×16) as well as implements a monopulse comparator. As in the previous work, and despite the higher complexity and size of this component, the measured results show great similarity with the simulation, which validates the printing direction orientation.

An assessment on planar and conformal 1D and 2D resonant SWAs is presented in [47]. Several orientations and machine settings were studied to obtain consistent parts with high detail resolution and quality. The study includes also investigation of manufacturing defects and surface roughness of the components.

Concerning non-resonant SWAs, it is worth highlighting the leaky-wave antennas [48] reported in [49–51], operating in different frequencies ranging from K- to V-band. Despite these publications involving plastic printing, the antenna in [50] was later used to build an array in metal. The novelty in these works is that additive manufacturing enables dual-polarization radiation from a single leaky-wave aperture and to consider an OMT printed together with the antenna. Moreover, inner ridges with modulated geometry have been considered inside the leaky waveguide, which enable low side-lobe level for the two linear polarizations. The array shown in **Figure 9** was conceived as an extension of the previous work and is here presented for the first time. The beamforming network is folded toward the back side of the antenna, and the whole is printed in a single piece. A summary of the measured performance is presented in **Figure 10**. A very good agreement with simulation is obtained for patterns and directivity over the operating bandwidth, demonstrating again the suitability of 3D-printing for SWA. The backward lobe in the pattern is produced by the short-circuit created at the end of the antenna. Such beam could be avoided by using a matched load.

Figure 9.
SLM printed compact ridged linearly polarized 2D leaky-wave array with folded beam-forming network.

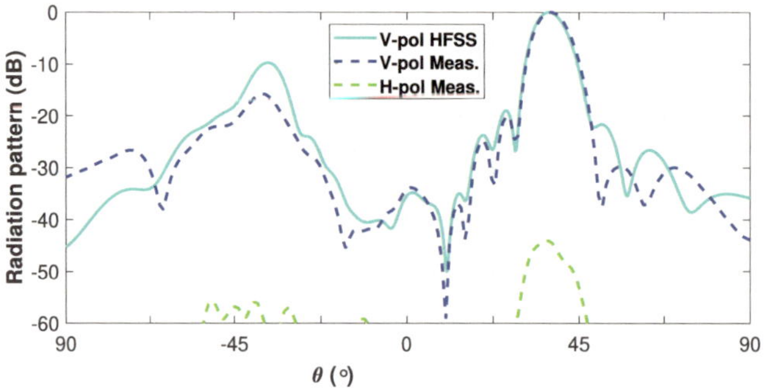

Figure 10.
E-plane pattern (dBi) of the SLM shorted array at 20 GHz.

Finally, an evolution of the previous works is presented in [52], which reports a fully metallic leaky-wave antenna radiating in circular polarization. One of the main original contributions of such work is related to the further plating of the SLM antenna. Such posttreatment step allows to reduce surface roughness and to improve the antenna radiation efficiency (around 10% improvement), as it can be seen in **Figure 11**.

5. Versatile periodic and quasi-periodic structures

Screens based on the periodic and quasi-periodic arrangement of unit cells are widely used nowadays due to their capability to alleviate the complexity of RF feeds in terms of operation frequency, beam shape, polarization, impedance matching, and focusing of the beam, among others. The screens that are currently employed in RF systems are mostly implemented in printed-circuit board (PCB) technology [53]. In order to allow for the wide industrialization of full-metal solutions, a change of paradigm is needed, both in RF design and in manufacturing.

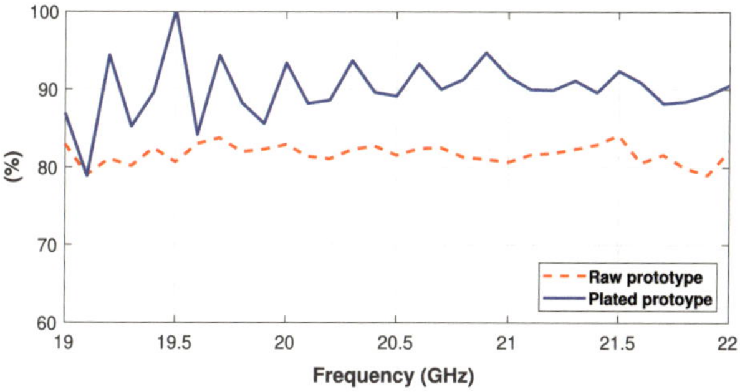

Figure 11.
Radiation efficiency (%) for both the raw and posttreated SLM circularly-polarized SWA in [52].

On the one hand, getting rid of the dielectric materials in the design of periodic surfaces is not evident, and it brings significant limitations to the RF designer. A monolithic architecture is needed that allows to be manufactured in metal without supports. A 3D topology is expected to be the future choice in this context, since it allows for greater design freedom. Additionally it is needed to count on a cell with sub-wavelength periodicity in order to avoid the appearance of grating lobes. Such an objective is easily attained in PCB solutions due to the miniaturization brought by the dielectric permittivity. However, when full-metal solutions are made sub-wavelength, they become very reactive and thus, tend to present a high reflection to the impinging waves. As a conclusion, a trade-off seems to appear in full-metal unit cells between the periodicity and the reflectivity when operating in transmission.

On the other hand, 3D-printing appears in this context as the technology for that enables greatest design freedom. However, the guidelines for metal 3D-printing have to be considered early in the phases of RF design.

Most of the metallic 3D screens that can be found in the literature are implemented through plastic 3D-printing and postprocessing of the piece with additional metal coating. This method can be effective, but it is very sensitive to such coating process. In addition, the effective conductivity of some commercial metallic inks or sprays can sometimes be lower than expected, contributing to a strong rise of ohmic losses. Some examples can be mentioned in this category. Negative-refractive index lenses have been reported in [54], metallic reflectarrays in [55, 56], and pass-band frequency selective surfaces in [57–59].

SLM has been applied for the design of periodic distributions of 3D helices, which would be very difficult to implement with other manufacturing techniques. Arrays of helices create artificial anysotropic panels that can be used for the design of polarization converters [60, 61]. Furthermore, additional dielectric supports were necessary to gain robustness. This fact dilutes the term "fully metallic" and strengthens the complexity of conceiving full-metal and self-supporting surfaces.

A metallic transmittarray has been reported in [62]. Its behavior is based on the excitation of a dispersive mode in a waveguide-shape unit cell. Solutions based on waveguide unit cells have also been proposed for the design of dual-band polarizers that provide orthogonal sense of polarization rotation in each band [63]. Such structure was based on highly subwavelength cells, in order to avoid excitation of grating lobes. Unfortunately, the previous designs were not driven by any co-design guideline. The previous geometries were conceived basing on classical RF guidelines, and they did not allow to be implemented in additive manufacturing.

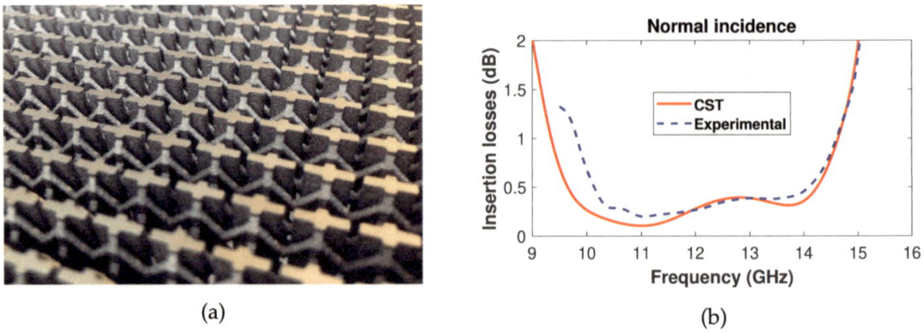

Figure 12.
3D-printed polarizing screen: (a) photograph of the prototype, (b) insertion losses (measured and simulated).

More recently, a topology of 3D unit cell has been proposed, which is specially suited to be 3D-printed [10, 64]. To the best knowledge of these authors, this is the first proof of concept in the literature of a periodic structure that is self-supporting and monolithically 3D-printed in metal. Such a screen is designed to achieve polarization conversion in a broad frequency band. The screen is low dispersive thanks to the excitation of transverse electromagnetic modes within the unit cell. A co-design approach based on the design guidelines detailed in Section 1 was followed in this case, thus enabling accurate prototyping. As an interesting example, bent metallic columns (inclined 45 with respect to the printing direction) were considered. The concept is illustrated in **Figure 12(a)**, which shows a Ku-band prototype printed in aluminum. The bent columns can be easily visualized in the photograph. The insertion losses are plotted in **Figure 12(b)**, they remain below 0.5 dB in the whole frequency band. These results are shown here for the first time. The axial-ratio level is quite similar to that reported in [10]. The architecture depicted in the picture can be adapted for other dual-polarization functionalities beyond polarization-conversion [64].

6. Conclusions

This chapter provides a wide overview of metal 3D-printing applied to the design of various types of waveguide components, antennas, and versatile periodic structures. It can be concluded that the use of 3D-printing can bring multiple relevant advantages when a set of guidelines are followed from early steps in the conception of the RF devices. This chapter opens wide perspectives for the conception of new disruptive RF devices through 3D-printing. The future trend concerns the development of 3D topologies with enhanced performance, complex functionality, and many degrees of freedom. New synergies are expected in RF design combined with powerful optimization techniques that are capable of managing such amount of geometrical parameters. Finally, new theoretical models are expected to appear, providing valuable physical insight on the phenomena underlying the operation of this new 3D structures.

Acknowledgements

This publication is supported by the European Union through the European Regional Development Fund (ERDF) and by the french region of Brittany, Ministry of

Higher Education and Research, Rennes Métropole and Conseil Départemental 35, through the CPER Project SOPHIE/STIC & Ondes, and also by the region of Brittany under grant SAD volet. Additional funding comes from the European Space Agency, the Centre National d'Études Spatiales, the companies Thales Alenia Space, Thales Group, and SWISSto12.

Conflict of interest

The authors declare no conflict of interest.

Author details

María García-Vigueras[1*], Lucas Polo-Lopez[1], Charalampos Stoumpos[1], Aurélie Dorlé[2], Carlos Molero[3] and Raphaël Gillard[1]

1 Institut d'Electronique et des Technologies du numéRique (IETR), INSA Rennes, Rennes, France

2 ONERA – The French Aerospace Lab, Toulouse, France

3 Universidad de Granada, Granada, Spain

*Address all correspondence to: maria.garcia-vigueras@insa-rennes.fr

IntechOpen

© 2022 The Author(s). Licensee IntechOpen. This chapter is distributed under the terms of the Creative Commons Attribution License (http://creativecommons.org/licenses/by/3.0), which permits unrestricted use, distribution, and reproduction in any medium, provided the original work is properly cited.

References

[1] Sorrentino R, Peverini OA. Additive manufacturing: A key enabling technology for next-generation microwave and millimeter-wave systems [point of view]. Proceedings of the IEEE. 2016;**104**(7):1362-1366

[2] Guo C, Shang X, Lancaster MJ, Xu J. A 3-d printed lightweight x-band waveguide filter based on spherical resonators. IEEE Microwave and Wireless Components Letters. 2015; **25**(7):442-444

[3] Huang G-L, Zhou S-G, Chio T-H, Yeo T-S. Fabrication of a high-efficiency waveguide antenna array via direct metal laser sintering. IEEE Antennas and Wireless Propagation Letters. 2016;**15**: 622-625

[4] Booth P, Gilmore J, Lluch EV, Harvey M. Enhancements to satellite feed chain performance, testing and lead-times using additive manufacturing. In: 2016 10th European Conference on Antennas and Propagation (EuCAP). 2016. pp. 1–5

[5] Peverini OA, Lumia M, Calignano F, Addamo G, Lorusso M, Ambrosio EP, et al. Selective laser melting manufacturing of microwave waveguide devices. Proceedings of the IEEE. 2017; **105**(4):620-631

[6] Le Sage GP. 3d printed waveguide slot array antennas. IEEE Access. 2016;**4**: 1258-1265

[7] Silva JS, García-Vigueras M, Debogović T, Costa JR, Fernandes CA, Mosig JR. Stereolithography-based antennas for satellite communications in ka-band. Proceedings of the IEEE. 2017; **105**(4):655-667

[8] Dimitriadis AI, Debogović T, Favre M, Billod M, Barloggio L, Ansermet J-P, et al. Polymer-based additive manufacturing of high-performance waveguide and antenna components. Proceedings of the IEEE. 2017;**105**(4):668-676

[9] Addamo G, Peverini OA, Manfredi D, Calignano F, Paonessa F, Virone G, et al. Additive manufacturing of Ka-band dual-polarization waveguide components. IEEE Transactions on Microwave Theory and Techniques. 2018;**66**(8):3589-3596

[10] Molero C, Legay H, Pierré T, García-Vigueras M. Broadband 3D-printed polarizer based on metallic transverse electro-magnetic unit-cells. IEEE Transactions on Antennas and Propagation. June 2022;**70**(6):4632-4644. DOI: 10.1109/TAP.2022.3145435

[11] Booth P. Additive manufactured bandpass filters at Ka-band. In: 2019 IEEE MTT-S International Microwave Workshop Series on Advanced Materials and Processes for RF and THz Applications (IMWS-AMP). 2019. pp. 7–9

[12] Polo-López L, Sirci S, Calleau A, Capdevila S, Toso G, Menargues E, et al. Vertically printable evanescent mode filters. IEEE Microwave and Wireless Components Letters. 2022. DOI: 10.1109/LMWC.2022.3187781

[13] An Z, Makdissy T, Vigueras MG, Vaudreuil S, Gillard R. A metal-only reflectarray made of 3d phoenix cells. In: 2022 16th European Conference on Antennas and Propagation (EuCAP). 2022. pp. 1–5

[14] Charles A, Elkaseer A, Thijs L, Hagenmeyer V, Scholz S. Effect of process parameters on the generated

surface roughness of down-facing surfaces in selective laser melting. Applied Sciences. 2019;**9**(6). Article number: 1256. [Online]. Available from: https://www.mdpi.com/about/announcements/784

[15] Peverini OA, Addamo G, Lumia M, Virone G, Calignano F, Lorusso M, et al. Additive manufacturing of Ku/K-band waveguide filters: A comparative analysis among selective-laser melting and stereo-lithography. IET Microwaves, Antennas & Propagation. 2017;**11**(6): 1936-1942

[16] Sirci S, Menargues E, Billod M. Space-qualified additive manufacturing and its application to active antenna harmonic filters. In: 2021 IEEE MTT-S International Microwave Filter Workshop (IMFW). 2021. pp. 239-242

[17] Sirci S, Menargues E, Berry S. Triangular combline filters conceived for additive manufacturing. In: 2021 IEEE MTT-S International Microwave Filter Workshop (IMFW). 2021. pp. 151-154

[18] Available from: https://www.eos.info/en/additive-manufacturing/3d-printing-metal/eos-metal-systems/eos-m 290

[19] Available from: https://www.eos.info/en/additive-manufacturing/3d-printing-metal/eos-metal-systems/eos-m 400

[20] Available from: https://www.airbus.com/en/newsroom/stories/2021-02-large-scale-3d-printing-goes-to-space-on-airbus-eurostar-neo satellites

[21] Available from: https://swissto12.com/a-large-batch-of-3d-printed-waveguide-solutions-has-been-delivered-to-thales-alenia-space-for-the-eutelsat-konnect-vhts program/

[22] Available from: https://news.satnews.com/2022/03/16/lockheed-martin-space-partners-with-swissto12-caes-for-advanced-3d-printed-phased-array antennas/

[23] Available from: https://ecss.nl/standard/ecss-q-st-70-80c-processing-and-quality-assurance-requirements-for-metallic-powder-bed-fusion-technologies-for-space-applications-30-july 2021/

[24] Kähkönen H, Proper S, Ala-Laurinaho J, Viikari V. Comparison of additively manufactured and machined antenna array performance at ka-band. IEEE Antennas and Wireless Propagation Letters. 2022;**21**(1):9-13

[25] Tomassoni C, Peverini OA, Venanzoni G, Addamo G, Paonessa F, Virone G. 3D printing of microwave and millimeter-wave filters: Additive manufacturing technologies applied in the development of high-performance filters with novel topologies. IEEE Microwave Magazine. 2020;**21**(6):24-45

[26] Wen X, Guo C, Shang X, Yu Y, Shu M, Yang Q, et al. SLM printed waveguide dual-mode filters with reduced sensitivity to fabrication imperfections. IEEE Microwave and Wireless Components Letters. 2021; **31**(11):1195-1198

[27] Uher J, Jens B, Uwe R. Artech House Publishers, 1993

[28] Booth P, Skeen M, Stirland S. Low cost, short lead-time feed chain components for multi-beam antennas. In: 2009 3rd European Conference on Antennas and Propagation. 2009, pp. 853-857

[29] Tribak A, Cano JL, Mediavilla A, Boussouis M. Octave bandwidth compact turnstile-based orthomode

transducer. IEEE Microwave and Wireless Components Letters. 2010; **20**(10):539-541

[30] Stoumpos C, Fraysse J-P, Goussetis G, Sauleau R, González CG, Pierré T. et al. Four-way orthomode waveguide power dividers: Subtractive and additive manufacturing. In: 2021 15th European Conference on Antennas and Propagation (EuCAP). 2021. pp. 1-5

[31] Stoumpos C, Fraysse J-P, Goussetis G, Sauleau R, Legay H. Quad-furcated profiled horn: The next generation highly efficient GEO antenna in additive manufacturing. IEEE Open Journal of Antennas and Propagation. 2022;**3**:69-82

[32] Kilian M, Schinagl-Weiß A, Kohl P, Sommer A, Hartwanger C, Schneider M. Additive layer manufactured waveguide RF components. In: 2019 49th European Microwave Conference (EuMC). 2019. pp. 790-793

[33] Addamo G, Peverini OA, Calignano F, Manfredi D, Paonessa F, Virone G, et al. 3-D printing of high-performance feed horns from Ku- to V-bands. IEEE Antennas and Wireless Propagation Letters. 2018;**17**(11):2036-2040

[34] Manafi S, Al-Tarifi M, Filipovic DS. 45–110 GHz quad-ridge horn with stable gain and symmetric beam. IEEE Transactions on Antennas and Propagation. 2017;**65**(9):4858-4863

[35] Cailloce Y, Hourlay P, Lebrun F, Palacin B. Additive manufacturing of Ku band horn antennas for telecommunications space applications. In: 12th European Conference on Antennas and Propagation (EuCAP 2018). 2018. pp. 1-4

[36] Huang G-L, Zhou S-G, Sim C-Y-D, Chio T-H, Yuan T. Lightweight perforated waveguide structure realized by 3-D printing for RF applications. IEEE Transactions on Antennas and Propagation. 2017;**65**(8):3897-3904

[37] Veljovic MJ, Skrivervik AK. Circularly polarized axially corrugated feed horn for CubeSat reflectarray applications. In: 2020 14th European Conference on Antennas and Propagation (EuCAP). 2020. pp. 1-4

[38] Foged LJ, Giacomini A, Morbidini R, Saccardi F, Schirosi V, Boumans M, et al. Investigation of additive manufacturing for broadband choked horns at X/Ku band. IEEE Antennas and Wireless Propagation Letters. 2018;**17**(11): 2003-2007

[39] Agnihotri I, Sharma SK. Design of a 3D metal printed axial corrugated horn antenna covering full Ka-band. IEEE Antennas and Wireless Propagation Letters. 2020;**19**(4):522-526

[40] Fenech H, Sonya A, Tomatis A, Soumpholphakdy V, Merino JLS. Eutelsat Quantum: A Game Changer. [Online]. DOI: 10.2514/6.2015-4318

[41] Elliott R, Kurtz L. The design of small slot arrays. IEEE Transactions on Antennas and Propagation. 1978;**26**(2): 214-219

[42] Tyagi Y, Mevada P, Chakrabarty S, Mahajan M. Broadband slotted waveguide array antenna with novel impedance matching network. In 2019 IEEE Indian Conference on Antennas and Propogation (InCAP). 2019. pp. 1-5

[43] Wang W, Zhong S-S, Zhang Y-M, Liang X-L. A broadband slotted ridge waveguide antenna array. IEEE Transactions on Antennas and Propagation. 2006;**54**(8):2416-2420

[44] Zhao H, Xu R, Wu W. Broadband waveguide slot array for SAR. Electronics Letters. 2011;**47**(02):76-77

[45] Tyagi Y, Mevada P, Chakrabarty S, Jyoti R. High efficiency broadband slotted waveguide array antenna. IET Microwaves, Antennas & Propagation. 2017;**11**:05

[46] Huang G-L, Zhou S-G, Chio T-H. Highly-efficient self-compact monopulse antenna system with integrated comparator network for RF industrial applications. IEEE Transactions on Industrial Electronics. 2017;**64**(1): 674-681

[47] Martin-Guennou A, Quéré Y, Rius E, Person C, Enouz-Vedrenne S, Lesueur G, et al. Direct metal laser sintering process investigation: Application to 3D slotted waveguide antennas. IET Microwaves, Antennas & Propagation. 2007;**11**(14): 1921-1929

[48] Oliner AA. In: Johnson RC, editor. Antenna Engineering Handbook. 3rd ed. New York: McGraw-Hill; 1993

[49] García-Vigueras M. Cost-effective dual-polarised leaky-wave antennas enabled by three-dimensional printing. IET Microwaves, Antennas Propagation. 2017;**11**(6):1985-1991

[50] Dorlé A, Gillard R, Menargues E, der Vorst MV, de Rijk E, Martín-Iglesias P et al. Sidelobe level reduction in ridged leaky waveguide through stereolithography. In: 2019 13th European Conference on Antennas and Propagation (EuCAP). 2019. pp. 1-5

[51] Dorlé A, Gillard R, Menargues E, Van Der Vorst M, De Rijk E, Martín-Iglesias P, et al. Additive manufacturing of modulated triple-ridge leaky-wave antenna. IEEE Antennas and Wireless Propagation Letters. 2018;**17**(11): 2123-2127

[52] Dorlé A, Gillard R, Menargues E, Van Der Vorst M, De Rijk E, Martín-Iglesias P, et al. Circularly polarized leaky-wave antenna based on a dual-mode hollow waveguide. IEEE Transactions on Antennas and Propagation. 2021;**69**(9):6010-6015

[53] Munk B. Frequency Selective Surfaces: Theory and Design. Melville, NY, USA: Wiley; 2005. Available from: https://aip.scitation.org/jap/info/contact

[54] Ehrenberg IM, Sarma SE, Wu B-I. A three-dimensional self-supporting low loss microwave lens with a negative refractive index. Journal of Applied Physics. 2012;**112**(7):073114

[55] Chen B-J, Yi H, Ng KB, Qu S-W, Chan, CH. 3D printed reflectarray antenna at 60 GHz. In: 2016 International Symposium on Antennas and Propagation (ISAP). 2016. pp. 92-93

[56] Velasco J, Parellada-Serrano I, Molero C. Fully metallic reflectarray for the Ku-band based on a 3D architecture. Electronics. 2021;**10**(21)

[57] Tang W, Zhu J, Wang C, Ge J, Yu Z, and Zhuang W. Waveguide 3-D FSSs by 3-D printing technique. In: 2016 International Conference on Electromagnetics in Advanced Applications (ICEAA). 2016. pp. 675-678

[58] Kien N, Hong I-P. Application of metaheuristic optimization algorithm and 3D printing technique in 3D bandpass frequency selective structure. Journal of Electrical Engineering & Technology. 2020;**15**:02

[59] Fernández Álvarez H, Cadman D, Goulas A, Gómez M, Engstrom D,

Vardaxoglou J, et al. 3D conformal bandpass millimeter-wave frequency selective surface with improved fields of view. Scientific Reports. 2021;**11**:06

[60] Wu S, Yachin VV, Shcherbinin VI, Tuz VR. Chiral metasurfaces formed by 3D-printed square helices: A flexible tool to manipulate wave polarization. Journal of Applied Physics. 2019;**126**(10):103101

[61] Wu G-B, Zeng Y-S, Chan KF, Qu S-W, Chan CH. 3-D printed circularly polarized modified fresnel lens operating at terahertz frequencies. IEEE Transactions on Antennas and Propagation. 2019;**67**(7):4429-4437

[62] Wang X, Cheng Y, Dong Y. Millimeter-wave dual-polarized metal transmitarray antenna with wide gain bandwidth. IEEE Antennas and Wireless Propagation Letters. 2022;**21**(2):381-385

[63] Molero C, Menargues E, Garcia-Vigueras M. All-metal 3-D frequency-selective surface with versatile dual-band polarization conversion. IEEE Transactions on Antennas and Propagation. 2020;(02): 1-1

[64] Bermúdez D, Molero C, Legay H, Palacin B, Menargues E, Gillard R, et al. Synthesis of versatile functionalities in arrangements of 3D unit-cells. In: 2021 IEEE Conference on Antenna Measurements Applications (CAMA). 2021. pp. 372-376

Chapter 7

Additive Manufacturing of Optical Waveguides

Yushi Chu, Liling Dong, Yanhua Luo, Jianzhong Zhang and Gang-Ding Peng

Abstract

Optical waveguides play an important role in both scientific research and industrial applications. Additive manufacturing (AM) or three-dimensional (3D)-printing technology has great potential to revolutionize manufacturing of optical waveguides. AM offers a great opportunity in developing optical waveguides demanding new material compositions and structure designs for functionalities needed in fast-evolving modern applications such as Internet of things (IoT). These demands have become so diverse and sophisticated that the traditional waveguide manufacturing cannot meet. In this chapter, we briefly introduce optical fibers one of the most common typical optical waveguides and present the process and perspective of optical fiber fabrication by AM technology.

Keywords: waveguides, 3D printing, waveguides fabrication, 3D-printing silica optical fibers

1. Introduction

Optical waveguides have achieved great success in information transmission in the past decades, mainly due to the ultralow loss, large capacity, high power, and excellent mechanical robustness. Optical fiber as one of the most useful optical waveguides plays an essential role in telecommunications and forms today's Internet backbone. In this chapter, optical fiber is briefly presented in Introduction part, then some AM technologies are focused, and finally, the fabrications of optical fibers based on AM technology are introduced including the fabrication process and perspective.

Optical fiber is a flexible, transparent fiber made of glass or plastic that acts as a light-transmitting tool. Optical fiber usually consists of a core surrounded by a transparent cladding and a coating in order, shown in **Figure 1a**. The refractive index of the core is higher than cladding, creating the waveguide structure to transmit light by total internal reflection (TIR) as demonstrated in **Figure 1b**. Charles K. Kao firstly promoted that the loss of optical fiber could be reduced by removed impurities and applied as the communication medium when he worked at ITTT Standard Telephones and Cables in 1966. This pioneering work made him earn the Nobel Prize in Physics in 2009 [1–3]. However, it was impossible to fabricate ultrapure silica as Kao mentioned due to the technical limitation at that time. Fortunately, the first low-loss (20 dB/km

IntechOpen

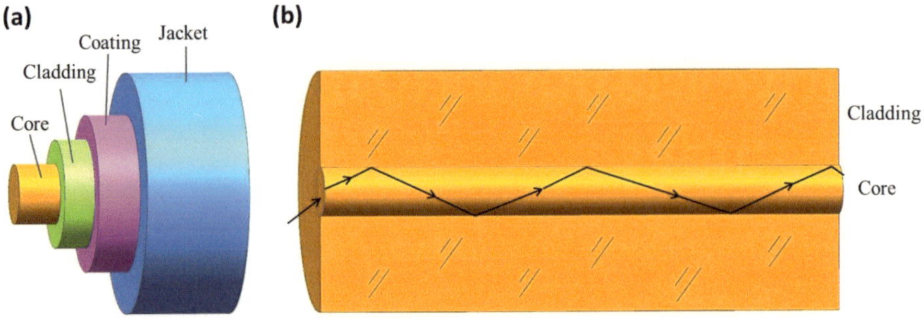

Figure 1.
(a) Diagram of typical optical fiber; (b) TIR in optical fiber.

at 632.8 nm) silica fiber was achieved by Robert D. Maurer from Corning in 1970 [4], and the modified chemical vapor deposition (MCVD) technology was invented by J. B. MacChesney from Bell Labs in 1974 [5]. Then, the optical fibers have developed rapidly and formed today's internet backbone. In 1999, Kao, Maurer, and MacChesney received the Charles Stark Draper Prize because of making the communication revolution possible [6].

Nowadays, there are some specialty optical fibers except the optical fiber for information transmission. The most representative ones are active fiber and microstructure optical fiber. For the active fiber, rare earth (RE) ions or metal ions are doped into optical fibers, generating luminescence under excitation, such as ytterbium (Yb) [7], erbium (Er) [8], thulium (Tm) [9], holmium (Ho) [10], and bismuth (Bi) [11]. Specific functions can also be achieved by codoping of two or more ions, for example, an ultra-broadband emission covering O-L telecommunication band was obtained from Bi/Er codoped optical fiber under 830-nm pumping, shown in **Figure 2a** [12, 13]. For the microstructure optical fiber, it usually consisted of one or more materials arranged periodically along the fiber length, realizing the refractive index modulation. The principles of light transmission are photonic bandgap effect and anti-resonance effect besides the TIR mentioned above. Microstructure optical fiber has many unique and novel physical properties, such as controllable nonlinearity, endless single-mode behavior, adjustable singular dispersion, low bending loss, and large mode field. **Figure 2b–e** shows the structures of typical PCFs [14–16].

The fabrication of optical fiber usually consists of two steps of preform manufacturing and fiber drawing, shown in **Figure 3**. The fiber drawing process is usually operated on a fiber drawing tower. The silica preform is heated to around 2000°C and becomes soft, then a thin bare fiber can be pulled out and cooled to solid, and finally, the bare fiber is coated and rolled into a coil, demonstrated in **Figure 3c**. For preform fabrication, chemical vapor deposition (CVD) is usually used for regular optical fiber, and microstructure optical fiber preform is manufactured by the stacking method. CVD utilizes $SiCl_4$ and $GeCl_4$ to oxidize into SiO_2 and GeO_2 at high temperature, which are deposited layer by layer onto the inner side of the quartz tube and sintered to form an optical fiber preform. According to the different deposition ways and heating source, CVD technologies can be divided into plasma chemical vapor deposition (PCVD) [17], outside vapor deposition (OVD) [18], vapor axial deposition (VAD) [19], and also MCVD [5] mentioned above. **Figure 3a** shows the schematic view of the four preform fabrication processes. Stack-and-draw method is usually used for PCF fabrication; glass capillaries

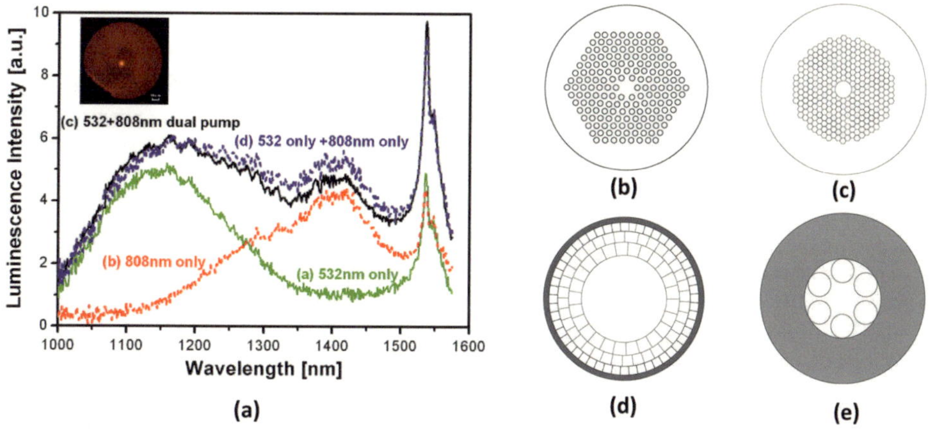

Figure 2.
(a) Emission spectra of Bi/Er codoped optical fibers [12]; (b)-(e) structures of large mode area fibers, hollow core fiber, Bragg fiber, and anti-resonance fiber.

and rods of a specific size are stacked according to the designed structure as shown in **Figure 3b** and then drawn into PCF [20].

However, as the Internet has evolved into the ubiquitous Internet of things (IoT), the role of optical fibers is expanding from passive telecommunications transmission medium to lasers, sensors, devices, and beyond. This is creating the demand for

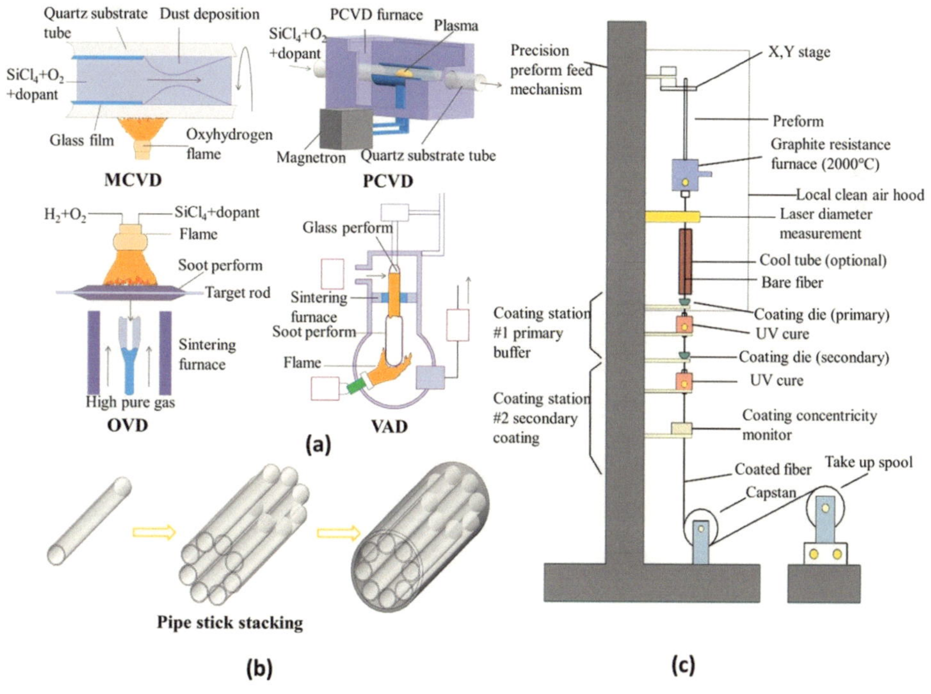

Figure 3.
Optical fiber fabrication process, (a) preform fabrication-based CVD technology, (b) preform fabrication-based stack technology, (c) fiber drawing process.

increasingly sophisticated optical fibers. Unfortunately, the traditional fabrication technologies, for example, the CVD mentioned above, have limited capability in both material and structure flexibility for diverse and custom-designed functionalities. AM technology provides a solution to this problem.

2. Additive manufacturing technology

AM technology is a kind of rapid prototyping technology, also known as 3D-printing technology. It is a technique of constructing objects by layer-by-layer printing using bondable materials such as metal powder or plastic, based on a 3D digital model. Compared with traditional manufacturing methods, AM technology has outstanding performance in terms of economic cost, time efficiency, and customized design and has been successfully applied to various materials such as metals, polymers, metamaterials, and composite materials. These materials are shaped using different principles such as sintering, melting, extrusion, and laser/ultraviolet light curing, exhibiting different manufacturing accuracies, printing speeds, and resolutions. According to different materials, AM technology can be divided into solid-based AM technology and liquid-based AM technology, shown in **Figure 4**. For the solid-based AM technology, powder and filament are manufactured to objects by laser sintering/melting (selective laser sintering, SLS; selective laser melting, SLM) and nozzle extrusion (fused deposition modeling, FDM). For liquid-based AM technology, ink or photosensitive resin is shaped by gelation (direct ink writing, DIW) and light polymerization (stereolithography, SLA; digital light processing, DLP; and polymer jet, Polyjet). The above technologies have made great contributions to AM technology waveguides, providing new possibilities for the diversity of waveguide structures and functions.

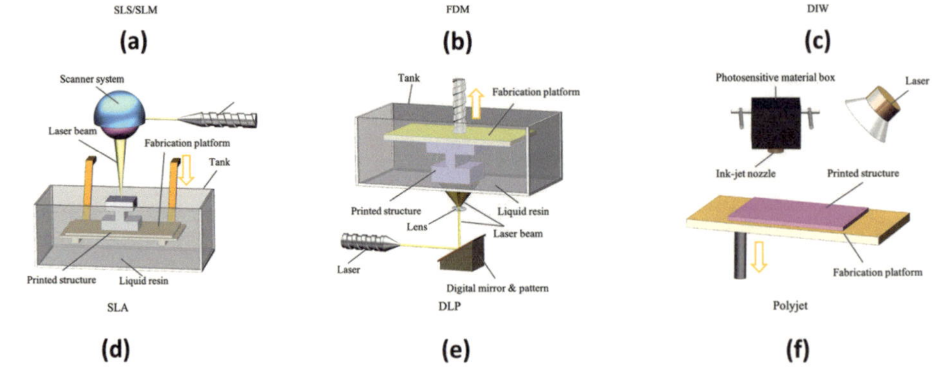

Figure 4.
Classification and schematic diagram of AM technology, (a) SLS/SLM, (b) FDM, (c) DIW, (d) SLA, (e) DLP, and (f) Polyjet.

2.1 Powder-based additive manufacturing technology

SLS/SLM technology is a solid powder-based AM technology that uses a laser to selectively scan the powder bed and utilizes the interaction of the laser with the solid powder to sinter/melt loose powders together as shown in **Figure 4a** [21]. The basic bonding mechanisms of SLS/SLM include solid-state sintering, chemically induced bonding, liquid-phase sintering-partial melting, and full melting [22]. The essence of solid-state sintering is the thermal diffusion of the melting temperature of the material between the particles through heat transfer, which can realize the processing and combination of a variety of materials. When the laser-powder interaction time is short and no binder is present, a chemically induced bonding mechanism occurs, forming a new binder phase that helps facilitate the post-curing process. Liquid-phase sintering-partial melting occurs when insufficient laser heat is provided and only partial melting is achieved to obtain the bonding of structural particles. Full melting is mainly used in metal materials, which can directly produce nearly full dense materials without post-processing.

Compared with other AM technologies, the main advantage of SLS/SLM is the flexibility of material selection. Single-component powder particles, composite powder particles, mixtures of different powder particles, and different binder materials are all suitable for SLS/SLM. The size and shape of the particles directly determine the shrinkage, precision, and density of the constructed object [23, 24]. There are van der Waals forces between particles of smaller size, and it is easy to form agglomerates, which may directly lead to uneven dispersion of powder particles; while larger-sized particles directly affect the porosity, showing poor surface roughness, even significant cracking or delamination effects may form [25]. The quality of the constructed object is highly dependent on the correct choice of the processing parameter setting, such as laser power, layer thickness, scan speed, as well as hatch spacing [26]. Uniform particle size distribution and optimized processing parameters can effectively reduce the occurrence of "step effect" and shrinkage deformation and finally obtain a satisfactory SLS/SLM construction object [27].

2.2 Filament-based additive manufacturing technology

FDM—the most famous AM technology, belongs to filament-based technology, which is also one of the most widely used technologies in rapid prototyping technology. FDM uses a form similar to squeezing toothpaste to extrude material to build a layered structure. By heating the filamentous thermoplastic material, the nozzle extrudes the material along the printing path under computer control and deposits it on the hotbed. After one layer, the hotbed moves up the distance of one layer of material thickness and then prints the second layer; layer-by-layer deposition finally realizes the overall printing, shown in **Figure 4b** [21]. The key to FDM is the temperature at which the nozzle is heated. At higher temperatures, the viscosity of the filamentous material decreases, resulting in reduced accuracy and even collapse and deformation during the printing process; the filamentous material does not melt completely at lower temperatures, which is prone to delamination and even blocks the nozzle [28].

FDM has the characteristics of simple process flow, low cost, and low environmental requirements. It can also realize multi-material printing by switching nozzles. However, FDM also has disadvantages, such as low-printing accuracy, poor mechanical strength, and being prone to the "step effect" [28]. When using FDM to build an object with a certain degree of complexity, the support of supporting materials is required to ensure the stability of the built part; otherwise, there may be collapse, which directly affects the manufacturing accuracy. In the actual construction process, the support structure can be printed with water-soluble filaments and then removed after post-processing, and the related problems caused by the support structure can be effectively reduced by optimizing the topology structure [29], optimizing the printing direction [30], and decomposing the printing structure to reduce the use of support materials [31].

2.3 Ink-based additive manufacturing technology

DIW technology is fabricated by layer-by-layer deposition, using an ink with shear-thinning properties that is extruded from a nozzle, shown in **Figure 4c**. The ink needs good viscoelasticity and fluidity to ensure smooth extrusion and maintain a good shape without clogging the nozzles. Because of the high accommodating capacity of ink materials, multi-component printing can be realized under the condition of meeting the basic requirements of ink, which can be widely used in many fields. Usually, the resolution of DIW technology is directly limited by the diameter and path of the nozzle. Latest studies have found that by adjusting the printing-related parameter set, it can successfully break through the constraints of line width and shape to achieve higher-resolution printing.

DIW technology has the advantages of simple process and multi-material printing, but there is also a disadvantage. Green bodies printed with DIW usually require long-term low-temperature drying to remove unnecessary solvents, so this step is critical to avoid cracks formed by drying.

2.4 Photosensitive resin-based additive manufacturing technology

SLA technology uses a laser with a specific wavelength to scan the photosensitive resin from point to line and to the surface according to the set path. After the layer is cured, move the worktable and apply a new layer of liquid resin on the surface of the original cured resin, so as to scan and cure the next layer, so that the newly cured layer can be firmly bonded to the previous layer. Repeating until the entire object has been fabricated, shown in **Figure 4d**. The key process parameters for SLA technology include initiator concentration, laser intensity, and scan rate. The initiator concentration and laser intensity directly determine the speed of the photopolymerization reaction, which in turn affects the processing efficiency. The spot size of the laser directly determines the resolution of the SLA [32]. Compared with FDM and SLS/SLM technologies, SLA shows better surface finish (nanoscale) and precision (micron scale), but the materials used are limited and it is difficult to achieve multi-material printing.

DLP technology is another AM technology that uses UV lamp curing. Its working principle is similar to SLA. The light source is changed from laser to UV lamp, and the introduction of digital micromirror devices (DMD) significantly shortens the construction time. The working principle of DLP technology is that the UV lamp is used as the light source, the three-dimensional CAD model is sliced through the software to form a two-dimensional dynamic mask pattern, and the sliced image is projected onto the surface of the photosensitive resin by DMD and cured, and the curing of one layer is completed, demonstrated in **Figure 4e**. The build plate is coated with a new layer of photosensitive resin, and the process is repeated layer by layer until the overall build is complete. DLP technology is one of the most widely used rapid prototyping technologies [33]. Compared with SLA technology, DLP technology has the characteristics of fast printing speed, high printing accuracy, and uniform mechanical strength [34]. The curing depth is the key to the success of the layer connection. The setting of the layer thickness should be appropriately smaller than the curing depth to ensure the connection between each layer while avoiding the shrinkage and deformation problems caused by excessive curing and eliminating anisotropy [35].

Polyjet technology also uses UV lamp curing. Its working principle is that the nozzle controlled by the X-/Y-axis sprays photosensitive resin on the printing platform, and each layer of material is sprayed and irradiated with UV light for photocuring. After each layer is cured, the printing platform drops along the Z-axis direction, and the previous process is repeated until the printing is finished, exhibited in **Figure 4f**. Like the FDM technology, a support structure is usually required. The support material and the structural material are sprayed at the same time to achieve stable support for complex or overhanging structures. The support materials are generally water-soluble materials, which are easy to remove later [36]. Polyjet technology has many advantages of rapid prototyping manufacturing, providing good surface finish, high precision and layer resolution, high printing efficiency, and also enabling multi-material printing.

3. Optical fiber fabrication by additive manufacturing technology

With the rapid development of modern technology and application fields, the requirements for optical fibers in terms of structure and materials are more complex and diverse, striving to achieve the customized structural design and free and flexible material combinations. However, traditional optical fiber manufacturing methods such as CVD or stack-and-draw are difficult to meet this demand. Optical fiber manufacturing based on CVD processes is limited by the lathe itself, which has limited structural flexibility, and it is difficult to realize the manufacture of complex optical fiber structures, and it is necessary to maintain a highly uniform radially symmetrical temperature and pressure during the manufacturing process [37]. Although the stack-and-draw can achieve a certain degree of structural complexity, the flexibility of materials is limited, and it is very time-consuming and labor-intensive [38]. AM technology provides huge opportunities for the development of new cylindrical fibers such as new structural fibers.

3.1 Polymer optical fiber fabricated by additive manufacturing technology

Polymer optical fibers (POFs) are ideal materials for short-range communications and are increasingly used due to their low cost and high elastic strain limit. POF is usually fabricated by extrusion or drilling method, until Cook et al. reported that POFs were fabricated using FDM 3D-printing technology in 2015 [39]. The commercially available 3D-printing filament consisting of a propriety polystyrene mixture containing styrene-butadiene-copolymer and polystyrene was used as the raw materials. An optical fiber geometry consisting of a solid core surrounded by six air holes was chosen as an example and shaped to preform by FDM 3D printer, as shown in **Figure 5a**. The preform was then annealed and drawn to fibers. The light-guiding images of fiber end faces at 630 nm and 515 nm, white-light output projected onto a screen, and the relative schematic setup are demonstrated in **Figure 5b**. The loss of the as-made POF was ~1.5 dB/cm, ~0.75 dB/cm, and ~ 1.51 dB/cm at 632 nm, 1064 nm, and 1550 nm, respectively.

Compared with the traditional POF manufacturing technology, the optical fiber preform manufactured by AM technology greatly simplifies the manufacturing process and saves a lot of time and labor costs. In addition, due to the integrated manufacturing, the waste of materials caused by drilling and other methods in the traditional manufacturing process is avoided. Despite the great advantages of using AM technology to manufacture preforms, the loss of manufacturing optical fibers is still high. These losses mainly come from the scattering caused by the gap between layers during the printing process, which can be reduced by an additional annealing process or optimized fiber drawing process [39].

After the first demonstration of POF by AM technology, various types of POFs fabricated by AM technology were reported, such as terahertz (THz) fiber, photonic bandgap fibers, anti-resonant fibers, Bragg fibers, step index fiber with two materials, and magnetically doped fiber with multi-materials by FMD, DLP, STL, and Polyjet methods [40–46].

3.2 Silica optical fiber fabricated by additive manufacturing technology

Although POFs realize low-cost AM technology and are widely used in short-distance communication, silica fiber plays an irreplaceable role in the long-distance transmission process. Besides, silica material has unparalleled optical transparency, excellent electrical insulation, chemical resistance, and thermal stability. How to break through the traditional silica fiber manufacturing method and realize the AM of silica fiber has become a problem worthy of further study. In 2016, F. Kotz et al. proposed a soft lithography method and successfully mixed silica nanoparticles with monomer solution, combined with light curing to prepare "liquid glass," which provided a new way for silica glass fabricated by AM technology [47]. After that, the combination of organic matter and silica has been widely used in the AM of silica glass. For example, in order to bridge the gap between fused silica glass and polymer processing,

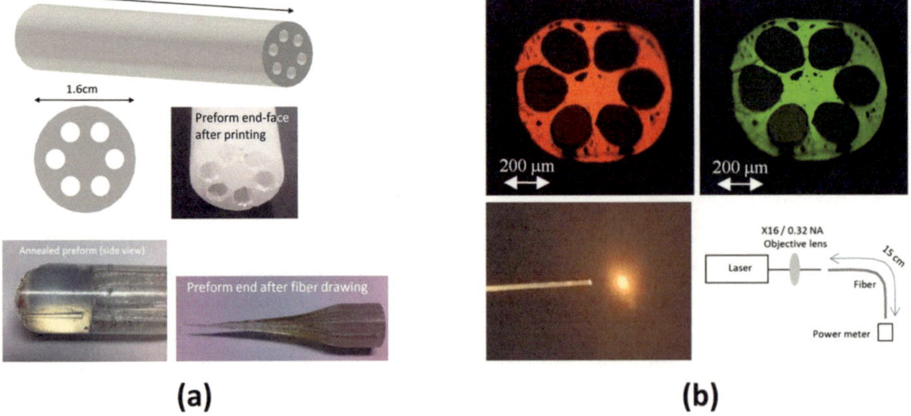

Figure 5.
(a) Illustration of POF preform design, photos of annealed preform and preform after fiber draw. (b) The light-guiding images of fiber end faces at 630 nm and 515 nm, white-light output projected onto a screen, and relative schematic setup.

"Glassomer" was introduced, which successfully combined the advantages of polymer processing with the excellent material properties of fused silica glass, which was heat treated to form the same optical properties as commercial fused silica glass [48]. Based on the previous two methods, the organic-inorganic hybrid photosensitive resin was prepared, and optical glass was successfully printed by DLP technology [49]. In addition, the use of DLP technology to induce the phase separation of the photosensitive resin provides the possibility for the fabrication of complex structures, and multi-component glass [50]. Lawrence Livermore National Laboratory (LLNL) has also developed a similar organic-inorganic hybrid silica material with sol-gel ink, silica glass printed by DIW technology, and optical glass doped with Ti and Ge [51–53].

3.2.1 Silica fiber fabrication based on DLP technology

The above-mentioned technologies for AM of silica glass lay the foundation for the AM of silica optical fibers. However, the size of silica glasses produced by AM was small, usually only 10 millimeters, which was far from the size of the optical fiber preforms. The main reason may be that the size of the preform is large, impurities are likely to remain in the preform after debinding, and the cooling rate after sintering is too slow, resulting in crystallization or ceramicization, and the overall devitrification of the preform. In response to this problem, Chu. Y et al. optimized the method and successfully used DLP AM technology to manufacture traditional single-mode fiber and multimode fiber in 2019 [54]. Then, the research group continued their work and fabricated multi-component and multicore optical fibers using DLP 3D-printing technology [55]. The process of bismuth and erbium codoped optical fiber (BEDF) manufactured by AM technology is shown in **Figure 6**.

Firstly, silica nanoparticles were dispersed into a UV photosensitive monomer forming a stable photosensitive resin. Secondly, the pre-designed optical fiber preform was printed by DLP printer, and functional materials such as Ge^{4+}, Er^{3+}, and Bi^{3+} were doped into the core. Ge^{4+} was used to adjust the refractive index to realize the waveguide structure, but Er^{3+} and Bi^{3+} were utilized to achieve the broadband near-infrared luminescence, presented in **Figure 6a–c**. Thirdly, the printed preform was moved to a furnace to remove the organic components and achieves densification. The temperature setting is shown in **Figure 6d**, the mass and size change clearly pointed out that the organics were removed before 600°C and the preform started to densify after 600°C. Finally, the preform was drawn to the fiber at 1855°C. Before drawing, the preform was heated to 810°C and kept for 3 hrs for removing the moisture absorbed by the preform due to the porous structure during storage.

The drawn fiber was characterized by X-ray diffraction (XRD) using the powdered BEDF without coating, and the pattern pointed out that the fiber was amorphous, shown in **Figure 6e** inset. The cross-sectional microscopy images and electron probe micro-analysis (EPMA) of BEDF are shown in **Figure 6f** and **Figure 6g**, respectively. Although the crack and shrinkage were noticed, multicore structures were kept. Elements distribution of BEDFs was as excepted, resulting in a refractive index difference between the core and cladding to form the waveguide structure, shown in **Figure 6g–i**. The loss and emission spectra of BEDF are demonstrated in **Figure 6j–k**. Typical characteristic peaks of bismuth and erbium were clearly identified.

DLP additive preform manufacturing has received attention from peers. In 2021, Zheng et al. have made progress in the AM of microstructured optical fibers; a ytterbium-doped microstructured optical fiber preform with a diameter of about 12 mm and a length of 20 mm was fabricated. The preform was then drawn to fibers,

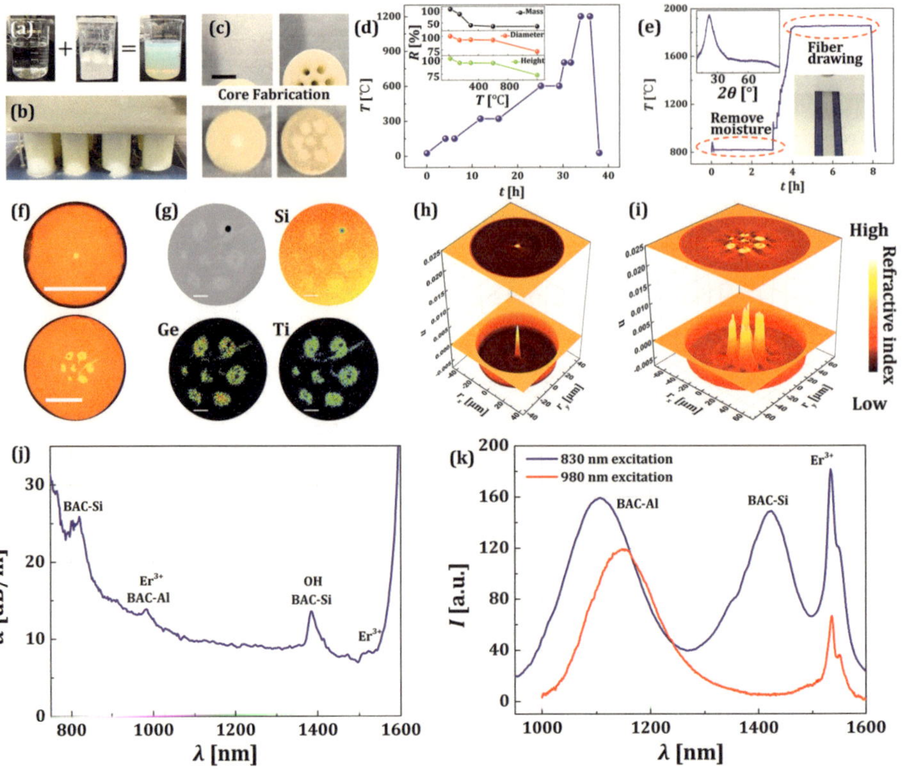

Figure 6.
(a) Dispersion of silica nanoparticle into ultraviolet curable resin, (b) preform cured by DLP 3D printing with UV light at 385 nm, (c) preforms before and after core filling, the scale bar is 10 mm, (d) temperature setting of the preform debinding process, inset is the remaining ratio of mass and size during the debinding process, (e) temperature change of the fiber drawing process. Insets are the XRD pattern of the drawn fiber and the photo of the drawing tower, (f) fiber cross-sectional images recorded by a microscope with a 50-μm scale bar, (g) cross-sectional view of seven-core BEDF, and EPMA-WDS mappings of different elements from the cross section (scale bar: 10 μm), (h)-(i) three-dimensional refractive index profiles of single- and seven-core BEDF, (j) loss spectrum of the single-core BEDF, (k) emission spectra of a single-core BEDF excited by the 830-nm and 980-nm lasers [54, 55].

the core was doped with 0.1 wt% ytterbium oxide, and six air holes were evenly distributed in the cladding [56]. The relevant information is shown in **Table 1**.

However, there are several difficulties in the process of printing silica fiber with DLP technology:

1. Due to the small particle size distribution of silica nanoparticles, clusters are prone to occur. It is necessary to ensure that the silica nanoparticles are evenly dispersed in the UV photosensitive resin to avoid scattering caused by particle accumulation; at the same time, it is necessary to ensure the stable suspension of silica nanoparticles for several weeks, to avoid the sedimentation of silica due to long-term placement under the influence of gravity, especially in the case of high solid content.

2. The correct selection of the printing parameter is the key to the AM process. The determination of layer thickness is closely related to the irradiation dose, $C_d = D_p \ln(E/E_c)$, where C_d is the depth of cure, E is the exposure energy, E_c is the critical exposure energy, and D_p is the depth of penetration [60]. A semi-empirical

formula suggests that the layer thickness should be slightly lower than the cure depth to ensure a close connection between the layers. In addition, attention should be paid to the balance between printing efficiency and layer thickness exposure time parameter selection while ensuring successful printing.

3. In the heat treatment process, the choice of heating rate and temperature is decisive. High transparency and compactness are guaranteed with all organics removed, and cracks during debinding and anisotropy due to non-uniform shrinkage are avoided as much as possible.

4. Controlling the fiber drawing parameters during the fiber drawing process can effectively avoid collapse and deformation, for example, the fiber drawing temperature directly determines the surface tension and viscosity of the fiber.

AM	Year	Preform		Fiber				References
		Type	Size (mm)	Type	Size (µm)	Materials	Loss (dB/m)	
DLP	2019	Solid	D: 25 L: 50–100	SM	d: 131 d_c: 4	M_{Clad}: SiO_2 M_{Core}: SiO_2-GeO_2-TiO_2	13.4@532 13.9@660 114@1550	[54]
			D: 25 L: 50–100	MM	d: 242 d_c: 14	M_{Clad}: SiO_2 M_{Core}: SiO_2-GeO_2-TiO_2	11@1300 5.8@1550	
	2021	Micro-Stru.	D: 12 L: 20	—	d: 110 d_c: -	M_{Clad}: SiO_2 M_{Core}: SiO_2: Yb	11@800 14@1100	[56]
	2022	Solid	D: 22 L: 40–100	SM	d: 80 d_c: 3.5	M_{Clad}: SiO_2 M_{Core}: SiO_2-GeO_2-TiO_2: Al/Bi/Er	9.6@1300 6@1550	[55]
			D: 22 L: 40–100	7 cores	d: 150 d_c: 3–11	M_{Clad}: SiO_2 M_{Core}: SiO_2-GeO_2-TiO_2: Al/Bi/Er	10.7@633	
DIW	2020	Solid	D: 2 L: 7	MM	d: 100 d_c: 40	M_{Clad}: Fluoride M_{Core}: SiO_2: Er	63@980 152@1535	[57]
SLS	2019	Solid	D: 12 L: -	MM	d: 200 d_c: -	M_{Clad}: SiO_2 M_{Core}: SiO_2	23@800 28@1100	[58]
		Micro-Stru.	D: 40 L: -	—	—	—	—	
	2020	Solid	D: 18 L: -	—	d: 150 d_c: 11	M_{Clad}: SiO_2 M_{Core}: SiO_2-GeO_2	8.32@800 24@1100	[59]
		Micro-Stru.	D: 38 L: -	—	—	—	—	

Table 1.
Additive manufacturing on silica optical fibers (Micro-Stru.: micro-structure, D: outer diameter of preform, L: length of preform, SM: single mode, MM: multimode, d: diameter of optical fiber, d_c: diameter of fiber core, M_{Clad}: material of fiber cladding, M_{Core}: material of fiber core).

3.2.2 Silica fiber fabrication based on DIW technology

At present, silica fiber preforms can also be made by DIW AM technology using inks mixed with silica and organics besides the DLP method. Erbium-doped optical fiber (EDF) was reported by the University of Southampton using DIW method [57]. Hydrophobic fumed silica and erbium chloride were doped in organics, such as tetraethylene glycol dimethyl ether and polydimethylsiloxane, to form the ink. Then, the ink was printed layer by layer with a printing speed of 40 mm/s at room temperature, and each layer was fixed at 500 μm. The printed rod was debinded and consolidated to glass. Finally, the glass rod was inserted into a fluorinated tube and drawn to EDF, and the fluorinated tube was used to provide the refractive index contrast, resulting in the waveguide structure. The typical absorption peaks of erbium were clearly observed at 980 nm and 1535 nm with the absorption of 62.98 dB/m and 151.49 dB/m, respectively. In addition, the overall absorption was still notable mainly due to the residual OH, PDMS, and impurities from the starting materials.

3.2.3 Silica fiber fabrication based on SLS technology

The use of SLS technology to print silica optical fibers fully utilizes the advantages of SLS technology in producing complex geometric shapes and multi-component printing. Optical fiber preforms with various structures and multi-components are formed by selectively scanning powder layers layer by layer with a CO_2 laser beam [58]. Microstructured fibers and anti-resonant fibers were printed by SLS technology. In order to evaluate the performance of the printed silica, a solid preform with 12 mm diameter was printed and inserted into a glass tube and drawn to fiber. A refractive index difference of $4x10^{-4}$ was achieved between the printed preform and glass tube. The attenuation of the optical fiber was measured in the range of 600–1150 nm, and the lowest attenuation was 23 dB/m around 800 nm. The high attenuation was mainly induced by the purity of the starting silica powder and parameters setting during the 3D-printing process.

After a series of SLS manufacturing technology optimization, photonic crystal fiber, anti-resonance fiber, and multicore Ge-doped optical fiber preforms were successfully fabricated by the same research group. The multicore Ge-doped optical fiber realizes a waveguide structure with a step index of refraction, and the lowest loss after the drawing reduced to 8.32 dB/m around 800 nm. Experiments have proved that the transmission loss can be effectively reduced by optimizing the powder characteristics and printing parameters [59].

4. Conclusions

In this chapter, we mainly introduced the fabrication of optical fibers using additive manufacturing technology and briefly presented the basic concepts of optical fibers and typical additive manufacturing techniques. Compared with traditional waveguide manufacturing techniques, such as CVD, additive manufacturing technology has higher manufacturing precision and manufacturing efficiency and lower manufacturing costs. At the same time, additive manufacturing technology exhibits unparalleled flexibility in structural and material designs that is essential to the realization of new and multifunctional optical fibers and devices.

At present, the additive manufacturing of polymer optical fiber preform mainly includes melt extrusion and light curing. Although the melt extrusion method has relatively low cost, the resolution of the manufacturing preform is low. While the light-curing technology is just the opposite. The manufacturing cost is higher than that of melt extrusion, and the resolution is also high, which is suitable for manufacturing preform with complex structures, such as bandgap optical fiber preform.

For the silica optical fiber preform, there are two main additive manufacturing methods. The first method is to combine silica and organic matter to form the resin, then use additive manufacturing technologies, such as light or heat curing, to shape, and finally obtain the optical fiber through debinding, sintering, and fiber drawing. This manufacturing method has a very high fabricating resolution and becomes the main way of additive manufacturing of silica optical fiber preform. However, due to the introduction of organic matter, an additional debinding process is required in the later stage, even if a small amount of organic matter remains, it will cause high loss. The second method is to directly sinter or melt the silica powder by laser. This method can effectively avoid the organic matter, but its development is also restricted by the problems of manufacturing accuracy and ceramics caused by phase transformation. The technologies above make full use of the advantages of additive manufacturing technology in molding, such as short time, low labor cost, and low material cost, and fully reflect its potential in the manufacturing of complex geometry silica optical fiber.

Although optical fiber fabricated by additive manufacturing has so many advantages, loss, manufacturing size, and multi-materials are still the factors limiting its development, and it is also the research direction in future. For the loss of additive manufacturing optical fiber, especially for the silica optical fiber, the loss mainly comes from microbubbles, microcracks, stripes between layers during printing, organic matter not removed during debinding process, and the purity of raw materials. For the manufacturing size, the main problems rise from the loss caused by incomplete removal of internal organic matter during debinding due to the large size of preform, and the cracking caused by uneven stress distribution during debinding, sintering, or cooling process. For the additive manufacturing of multi-material optical fiber, the integration of glass, semiconductor, crystal, metal, or polymer into the so-called hybrid fiber is also another research focus of additive manufacturing of fibers, and the key point is how to balance the relationship between melting point and thermal expansion coefficient of each material. At present, a large number of researchers have carried out systematic research on the above problems. We have reason to believe that just like the development trend of traditional optical fiber, the optical fiber fabricated by additive manufacturing will also experience the development trend of reducing loss, multi-structure, and multi-material and bring revolutionary changes to the optical fiber manufacturing industry with its unparalleled advantages.

Author details

Yushi Chu[1,2], Liling Dong[1,2], Yanhua Luo[3], Jianzhong Zhang[1*] and Gang-Ding Peng[3]

1 Key Laboratory of In-fiber Integrated Optics of Ministry of Education, College of Physics and Optoelectronic Engineering, Harbin Engineering University, Harbin, China

2 Fiber Optical Sensing Center for Excellence, Yantai Research Institute, Harbin Engineering University, Yantai, China

3 Photonics and Optical Communications, School of Electrical Engineering and Telecommunications, University of New South Wales, Sydney, NSW, Australia

*Address all correspondence to: zhangjianzhong@hrbeu.edu.cn

IntechOpen

© 2022 The Author(s). Licensee IntechOpen. This chapter is distributed under the terms of the Creative Commons Attribution License (http://creativecommons.org/licenses/by/3.0), which permits unrestricted use, distribution, and reproduction in any medium, provided the original work is properly cited.

References

[1] Kao KC, Hockham GA. Dielectric-fibre surface waveguides for optical frequencies. Proceedings of the Institution of Electrical Engineers. 1966;**113**:1151-1158. DOI: 10.1049/piee.1966.0189

[2] Kao KC. Nobel Lecture: Sand from centuries past: Send future voices fast. Reviews of Modern Physics. 2010;**82**:2299. DOI: 10.1103/RevModPhys.82.2299

[3] Press release. NobelPrize.org. [Internet]. 2009. Available from: https://www.nobelprize.org/prizes/physics/2009/press-release/

[4] Kapron FP, Donald BK, Maurer RD. Radiation losses in glass optical waveguides. Applied Physics Letters. 1970;**17**:423-425. DOI: 10.1063/1.1653255

[5] MacChesney JB, O'connor PB, Presby HM. A new technique for the preparation of low-loss and graded-index optical fibers. Proceedings of the IEEE. 1974;**62**:1280-1281. DOI: 10.1109/PROC.1974.9608

[6] Recipients of the Charles Stark Draper Prize [Internet]. 1999. Available from: https://www.nae.edu/55048/page1999

[7] Zhou P, Wang X, Xiao H, Ma Y, Chen J. Review on recent progress on Yb-doped fiber laser in a variety of oscillation spectral ranges. Laser Physics. 2012;**22**:823-831. DOI: 10.1134/S1054660X12050404

[8] Bradley JDB, Pollnau M. Erbium-doped integrated waveguide amplifiers and lasers. Laser & Photonics Reviews. 2011;**5**:368-403. DOI: 10.1002/lpor.201000015

[9] Sincore A, Bradford JD, Cook J, Shah L, Richardson MC. High average power thulium-doped silica fiber lasers: Review of systems and concepts. IEEE Journal of Selected Topics in Quantum Electronics. 2017;**24**:1-8. DOI: 10.1109/JSTQE.2017.2775964

[10] Zhou P, Wang X, Ma Y, Lv H, Liu Z. Review on recent progress on mid-infrared fiber lasers. Laser Physics. 2012;**22**:1744-1751. DOI:10.1134/s1054660x12110199

[11] Dianov EM. Bismuth-doped optical fibers: A challenging active medium for near-IR lasers and optical amplifiers. Light: Science & Applications. 2012;**1**:e12-e12. DOI: 10.1038/lsa.2012.12

[12] Luo Y, Wen J, Zhang J, Canning J, Peng GD. Bismuth and erbium codoped optical fiber with ultrabroadband luminescence across O-, E-, S-, C-, and L-bands. Optics Letters. 2012;**37**:3447-3449. DOI: 10.1364/OL.37.003447

[13] Luo Y, Yan B, Zhang J, Wen J, He J, Peng GD. Development of Bi/Er co-doped optical fibers for ultra-broadband photonic applications. Frontiers of Optoelectronics. 2018;**11**:37-52. DOI: 10.1007/s12200-017-0764-y

[14] Xiao L, Demokan MS, Jin W, Wang Y, Zhao CL. Fusion splicing photonic crystal fibers and conventional single-mode fibers: Microhole collapse effect. Journal of Lightwave Technology. 2007;**25**:3563-3574. DOI: 10.1109/JLT.2007.907787

[15] Vienne G, Xu Y, Jakobsen C, Deyerl HJ, Jensen JB, Sørensen T, et al. Ultra-large bandwidth hollow-core guiding in all-silica Bragg fibers with nano-supports. Optics Express.

2004;**12**:3500-3508. DOI: 10.1364/OPEX.12.003500

[16] Newkirk AV, Lopez JEA, Correa RA, Schülzgen A, Mazurowski J. Anti-resonant hollow core fiber for precision timing applications. In: Astronomical Optics: Design, Manufacture, and Test of Space and Ground Systems. International Society for Optics and Photonics. San Diego, California, United States: SPIE; 2017. p. 104010F

[17] Geittner P, Kuppers D, Lydtin H. Low-loss optical fibers prepared by plasma-activated chemical vapor deposition (CVD). Applied Physics Letters. 1976;**28**:645-646. DOI: 10.1063/1.88608

[18] Schultz PC. Fabrication of optical waveguides by the outside vapor deposition process. Proceedings of the IEEE. 1980;**68**:1187-1190. DOI: 10.1109/PROC.1980.11828

[19] Izawa T, Sudo S, Hanawa F. Continuous fabrication process for high-silica fiber preforms. IEICE Transactions. 1979;**62**:779-785. DOI: naid/10010177421

[20] Knight JC, Birks TA, Russell PSJ, Atkin DM. All-silica single-mode optical fiber with photonic crystal cladding. Optics Letters. 1996;**21**:1547-1549. DOI: 10.1364/OL.21.001547

[21] Layani M, Wang X, Magdassi S. Novel materials for 3D printing by photopolymerization. Advanced Materials. 2018;**30**(41): 1706344. DOI: 10.1002/adma.20170 6344

[22] Kruth JP, Mercelis P, Van Vaerenbergh J, Froyen L, Rombouts M. Binding mechanisms in selective laser sintering and selective laser melting. Rapid Prototyping Journal. 2005;**11**:26-36. DOI: 10.1108/13552540510573365

[23] Tang Y, Futh JYH, Loh HT, Wong YS, Lu L. Direct laser sintering of a silica sand. Materials and Design. 2003;**24**:623-629. DOI: 10.1016/S0261-3069(03)00126-2

[24] Klocke F, Wirtz H. Selective laser sintering of zirconium silicate. Proceedings of Solid Freeform Fabrication Symposium. 1998:605-612. DOI: 10.26153/tsw/652

[25] Klocke F, McClung A, Ader C. Direct laser sintering of borosilicate glass. Proceedings of Solid Freeform Fabrication Symposium. 2004:214-219. DOI: 10.26153/tsw/6986

[26] Kruth JP, Wang X, Laoui T, Froyen L. Lasers and materials in selective laser sintering. Assembly Automation. 2003;**23**:357-371. DOI: 10.1108/01445150310698652

[27] Song JL, Li YT, Deng QL, Hu DJ. Rapid prototyping manufacturing of silica sand patterns based on selective laser sintering. Journal of Materials Processing Technology. 2007;**187-188**:614-618. DOI: 10.1016/j.jmatprotec.2006.11.108

[28] Klein J, Stern M, Franchin G, Kayser M, Inamura C, Dave S, et al. Additive manufacturing of optically transparent glass. 3D Printing and Additive Manufacturing. 2015;**2**:92-105. DOI: 10.1089/3dp.2015.0021

[29] Gaynor AT, Guest JK. Topology optimization considering overhang constraints: Eliminating sacrificial support material in additive manufacturing through design. Structural and Multidisciplinary Optimization. 2016;**54**:1157-1172. DOI: 10.1007/s00158-016-1551-x

[30] Zwier MP, Wits WW. Design for additive manufacturing: Automated

build orientation selection and optimization. Procedia CIRP. 2016;**55**:128-133. DOI: 10.1016/j.procir.2016.08.040

[31] Yu EA, Yeom J, Tutum CC, Vouge E, Miikkulainen R. Evolutionary decomposition for 3D printing. GECCO' 17: Proceedings of the Genetic and Evolutionary Computation Conference. 2017:1272-1279. DOI: 10.1145/3071178.3071310

[32] Cruz A, Cordeiro C, Franco M. 3D printed hollow-core terahertz fibers. Fibers. 2018;**6**(3):43. DOI: 10.3390/fib6030043

[33] Sung C, Fang N, Wu DM, Zhang X. Projection micro-stereolithography using digital micro-mirror dynamic mask. Sensors and Actuators A: Physical. 2005;**121**:113-120. DOI: 10.1016/j.sna.2004.12.011

[34] Wu D, Zhao Z, Zhang Q, Qi HJ, Fang D. Mechanics of shape distortion of DLP 3D printed structures during UV post-curing. Soft Matter. 2019;**15**:6151-6159. DOI: 10.1039/c9sm00725c

[35] Komissarenko D, Sokolov P, Evstigneeva A, Shmeleva I, Dosovitsky A. Rheological and curing behavior of acrylate-based suspensions for the DLP 3D printing of complex zirconia parts. Materials. 2018;**11**:2350. DOI: 10.3390/ma11122350

[36] Meisel NA, Gaynor A, Williams CB, Guest JK. Multiple-material topology optimization of compliant mechanisms created via polyjet 3d printing. In: Proceedings of the 24th Annual international solid freeform fabrication symposium an additive manufacturing conference; University of Texas at Austin; 2013; p. 980-997. DOI: 10.26153/tsw/15651

[37] Peng GD, Luo Y, Zhang J, Wen J, Chu Y, Cook K, et al. 3D silica lithography for future optical fiber fabrication. In: Handbook of Optical Fibers. Singapore: Springer Singapore; 2019. pp. 637-653. DOI: 10.1007/978-981-10-7087-7_79

[38] Anuszkiewicz A, Kasztelanic R, Filipkowski A, Stepniewski G, Stefaniuk T, Siwicki B, et al. Fused silica optical fibers with graded index nanostructured core. Scientific Reports. 2018;**8**:12329. DOI: 10.1038/s41598-018-30284-1

[39] Cook K, Canning J, Leon-Saval S, Reid Z, Hossain MA, Comatti JE, et al. Air-structured optical fiber drawn from a 3D-printed preform. Optics Letters. 2015;**40**:3966-3969. DOI: 10.1364/ol.40.003966

[40] Li J, Nallappan K, Guerboukha H, Skorobogatiy M. 3D printed hollow core terahertz Bragg waveguides with defect layers for surface sensing applications. Optics Express. 2017;**25**:4126-4144. DOI: 10.1364/oe.25.004126

[41] Yang J, Zhao J, Gong C, Tian H, Sun L, Chen P, et al. 3D printed low-loss THz waveguide based on Kagome photonic crystal structure. Optics Express. 2016;**24**:22454-22460. DOI: 10.1364/oe.24.022454

[42] Van Putten LD, Gorecki J, Numkam Fokoua E, Apostolopoulos V, Poletti F. 3D-printed polymer antiresonant waveguides for short-reach terahertz applications. Applied Optics. 2018;**57**:3953-3958. DOI: 10.1364/ao.57.003953

[43] Li J, Ma T, Nallapan K, Guerboukha H, Skorobogatiy M. 3D printed hollow core terahertz Bragg waveguides with defect layers for surface sensing applications. In: 2017 42nd International Conference on

Infrared, Millimeter, and Terahertz Waves (IRMMW-THz). IEEE: Cancun, Mexico; 2017. DOI: 10.1109/irmmw-thz.2017.8066908

[44] Cook K, Balle G, Canning J, Chartier L, Athanaze T, Hossain MA, et al. Step-index optical fiber drawn from 3D printed preforms. Optics Letters. 2016;**41**:4554-4557. DOI: 10.1364/ol.41.004554

[45] Kaufman JJ, Bow C, Tan FA, Cole AM, Abouraddy AF. 3D printing preforms for fiber drawing and structured functional particle production. In: Proceedings of the Australian Conference on Optical Fibre Technology. Optical Society of America. 2016. DOI: 10.1364/ACOFT.2016.AW4C.1

[46] Hong B, Swithenbank M, Greenall N, Clarke RG, Chudpooti N, Akkaraekthalin P, et al. Low-loss asymptotically single-mode THz Bragg fiber fabricated by digital light processing rapid prototyping. IEEE Transactions on Terahertz Science and Technology. 2018;**8**:90-99. DOI: 10.1109/tthz.2017.2778047

[47] Kotz F, Plewa K, Bauer W, Schneider N, Keller N, Nargang T, et al. Liquid glass: A facile soft replication method for structuring glass. Advanced Materials. 2016;**28**:4646-4650. DOI: 10.1002/adma.201506089

[48] Kotz F, Schneider N, Striegel A, Wolfschläger A, Keller N, Worgull M, et al. Glassomer—Processing fused silica glass like a polymer. Advanced Materials. 2018;**30**:1707100. DOI: 10.1002/adma.201707100

[49] Kotz F, Arnold K, Bauer W, Schild D, Keller N, Sachsenheimer K, et al. Three-dimensional printing of transparent fused silica glass. Nature. 2017;**544**:337-339. DOI: 10.1038/nature22061

[50] Moore DG, Barbera L, Masania K, Studart AR. Three-dimensional printing of multicomponent glasses using phase-separating resins. Nature Materials. 2019;**19**(2):212-217. DOI: 10.1038/s41563-019-0525-y

[51] Nguyen DT, Meyers C, Yee TD, Dudukovic NA, Destino JF, Zhu C, et al. 3D-printed transparent glass. Advanced Materials. 2017;**29**:1701181. DOI: 10.1002/adma.201701181

[52] Destino JF, Dudukovic NA, Johnson MA, Nguyen DT, Yee TD, Egan GC, et al. 3D printed optical quality silica and silica-titania glasses from sol-gel feedstocks. Advanced Materials Technologies. 2018;**3**:1700323. DOI: 10.1002/admt.201700323

[53] Sasan K, Lange A, Yee TD, Dudukovic NA, Nguyen DT, Johnson MA, et al. Additive manufacturing of optical quality germania–silica glasses. ACS Applied Materials & Interfaces. 2020;**12**:6736-6741. DOI: 10.1021/acsami.9b21136

[54] Chu Y, Fu X, Luo Y, Canning J, Tian Y, Cook K, et al. Silica optical fiber drawn from 3D printed preforms. Optics Letters. 2019;**44**:5358-5361. DOI: 10.1364/OL.44.005358

[55] Chu Y, Fu X, Luo Y, Canning J, Wang J, Ren J, et al. Additive manufacturing fiber preforms for structured silica fibers with bismuth and erbium dopants. Light. Advanced Manufacturing. 2022;**3**:1-7. DOI: 10.37188/lam.2022.021

[56] Zheng B, Yang J, Qi F, Wang J, Zhang X, Wang P. Fabrication of Yb-doped silica micro-structured optical fibers from UV-curable nano-composites and their application in temperature sensing. Journal of Non-Crystalline Solids. 2021;**573**:121129. DOI: 10.1016/j.jnoncrysol.2021.121129

[57] Rosales ALC, Núñez-Velázquez MMA, Sahu JK. 3D printed Er-doped silica fibre by direct ink writing. Proceedings of the EPJ Web of Conferences. 2020;**243**:20002. DOI: 10.1051/epjconf/202024320002

[58] Camacho-Rosales A, Núñez-Velázquez MMA, Zhao X, Yang S, Sahu JK. Development of 3-D printed silica preforms. In: 2019 Conference on Lasers and Electro-Optics Europe & European Quantum Electronics Conference (CLEO/Europe-EQEC). 2019. DOI: 10.1109/cleoe-eqec.2019.8871436

[59] Rosales ALC, Núñez-Velázquez MMA, Zhao X, Sahu JK. Optical fibers fabricated from 3D printed silica preforms. Proceedings of the Laser 3D Manufacturing VII. 2020;**11271**:112710U. DOI: 10.1117/12.2543210

[60] Liu C, Qian B, Liu X, Tong L, Qiu J. Additive manufacturing of silica glass using laser stereolithography with a top-down approach and fast debinding. RSC Advances. 2018;**8**:16344-16348. DOI: 10.1039/c8ra02428f